Steroids Made It Possible

PROFILES, PATHWAYS, AND DREAMS

Autobiographies of Eminent Chemists

Jeffrey I. Seeman, Series Editor

Steroids Made It Possible

Carl Djerassi

American Chemical Society, Washington, DC 1990

Library of Congress Cataloging-in-Publication Data

Djerassi, Carl.
  Steroids made it possible/Carl Djerassi.
    p. cm.—(Profiles, pathways, and dreams: autobiographies of eminent chemists, ISSN 1047–8329)

  Includes bibliographical references (p.   ) and index.

  ISBN 0–8412–1773–4 (cloth).—ISBN 0–8412–1799–8 (paper)

  1. Djerassi, Carl. 2. Chemists—United States—Biography. 3. Chemistry, Organic—History—20th century. 4. Steroids.

  I. Title. II. Series: Profiles, pathways, and dreams.

QD22.D63A3 1990
540'.92—dc20   90–906
[B]   CIP

The paper used in this publication meets the minimum requirements of American National Standard for Information Sciences—Permanence of Paper for Printed Library Materials, ANSI Z39.48–1984.

∞

Copyright © 1990

American Chemical Society

All Rights Reserved. The copyright owner consents that reprographic copies may be made for personal or internal use or for the personal or internal use of specific clients. This consent is given on the condition, however, that the copier pay the stated per-copy fee through the Copyright Clearance Center, Inc., 27 Congress Street, Salem, MA 01970, for copying beyond that permitted by Sections 107 or 108 of the U.S. Copyright Law. This consent does not extend to copying or transmission by any means—graphic or electronic—for any other purpose, such as for general distribution, for advertising or promotional purposes, for creating a new collective work, for resale, or for information storage and retrieval systems. The copying fee is $0.75 per page. Please report your copying to the Copyright Clearance Center with this code: 1047–8329/90/$00.00+.75.

The citation of trade names and/or names of manufacturers in this publication is not to be construed as an endorsement or as approval by ACS of the commercial products or services referenced herein; nor should the mere reference herein to any drawing, specification, chemical process, or other data be regarded as a license or as a conveyance of any right or permission to the holder, reader, or any other person or corporation, to manufacture, reproduce, use, or sell any patented invention or copyrighted work that may in any way be related thereto. Registered names, trademarks, etc., used in this publication, even without specific indication thereof, are not to be considered unprotected by law.

PRINTED IN THE UNITED STATES OF AMERICA

1990 ACS Books Advisory Board

V. Dean Adams
Tennessee Technological University

Paul S. Anderson
Merck Sharp & Dohme Research Laboratories

Alexis T. Bell
University of California—Berkeley

Malcolm H. Chisholm
Indiana University

Natalie Foster
Lehigh University

G. Wayne Ivie
U.S. Department of Agriculture, Agricultural Research Service

Mary A. Kaiser
E. I. du Pont de Nemours and Company

Michael R. Ladisch
Purdue University

John L. Massingill
Dow Chemical Company

Robert McGorrin
Kraft General Foods

Daniel M. Quinn
University of Iowa

Elsa Reichmanis
AT&T Bell Laboratories

C. M. Roland
U.S. Naval Research Laboratory

Stephen A. Szabo
Conoco Inc.

Wendy A. Warr
Imperial Chemical Industries

Robert A. Weiss
University of Connecticut

Foreword

In 1986, the ACS Books Department accepted for publication a collection of autobiographies of organic chemists, to be published in a single volume. However, the authors were much more prolific than the project's editor, Jeffrey I. Seeman, had anticipated, and under his guidance and encouragement, the project took on a life of its own. The original volume evolved into 22 volumes, and the first volume of *Profiles, Pathways, and Dreams: Autobiographies of Eminent Chemists* was published in 1990. Unlike the original volume, the series was structured to include chemical scientists in all specialties, not just organic chemistry. Our hope is that those who know the authors will be confirmed in their admiration for them, and that those who do not know them will find these eminent scientists a source of inspiration and encouragement, not only in any scientific endeavors, but also in life.

M. Joan Comstock
Head, Books Department
American Chemical Society

Contributors

We thank the following corporations and Herchel Smith for their generous financial support of the series Profiles, Pathways, and Dreams.

Akzo nv

Bachem Inc.

E. I. du Pont de Nemours and Company

Duphar B.V.

Eisai Co., Ltd.

Fujisawa Pharmaceutical Co., Ltd.

Hoechst Celanese Corporation

Imperial Chemical Industries PLC

Kao Corporation

Mitsui Petrochemical Industries, Ltd.

The NutraSweet Company

Organon International B.V.

Pergamon Press PLC

Pfizer Inc.

Philip Morris

Quest International

Sandoz Pharmaceuticals Corporation

Sankyo Company, Ltd.

Schering–Plough Corporation

Shionogi Research Laboratories, Shionogi & Co., Ltd.

Herchel Smith

Suntory Institute for Bioorganic Research

Takasago International Corporation

Takeda Chemical Industries, Ltd.

Unilever Research U.S., Inc.

About the Editor

JEFFREY I. SEEMAN received his B.S. with high honors in 1967 from the Stevens Institute of Technology in Hoboken, New Jersey, and his Ph.D. in organic chemistry in 1971 from the University of California, Berkeley. Following a two-year staff fellowship at the Laboratory of Chemical Physics of the National Institutes of Health in Bethesda, Maryland, he joined the Philip Morris Research Center in Richmond, Virginia, where he is currently a senior scientist and project leader. In 1983–1984, he enjoyed a sabbatical year at the Dyson Perrins Laboratory in Oxford, England, and claims to have visited more than 90% of the castles in England, Wales, and Scotland.

Seeman's 80 published papers include research in the areas of photochemistry, nicotine and tobacco alkaloid chemistry and synthesis, conformational analysis, pyrolysis chemistry, organotransition metal chemistry, the use of cyclodextrins for chiral recognition, and structure–activity relationships in olfaction. He was a plenary lecturer at the Eighth IUPAC Conference on Physical Organic Chemistry held in Tokyo in 1986 and has been an invited lecturer at numerous scientific meetings and universities. Currently, Seeman serves on the Petroleum Research Fund Advisory Board. He continues to count Nero Wolfe and Archie Goodwin among his best friends.

Contents

List of Photographs xiii

Preface xvii

Editor's Note xxi

Steroids Made It Possible 1

    Diary Entry (11 August 1983) 3

    The Partial Synthesis of Steroids 17

    Optical Rotatory Dispersion and Circular Dichroism 53

    Organic Mass Spectrometry 77

    Magnetic Circular Dichroism 100

    Applications of Computer Artificial Intelligence Techniques 102

    Marine Sterols and Phospholipids 114

    Epilogue 138

    Coda 161

References 171

Index 191

Photographs

With W. S. Johnson and G. Stork during a symposium at Stanford	xxiii
My father, Samuel Djerassi, and I in 1939 in Sofia	4
My mother, Alice Djerassi, and I in 1939	5
Paul Meier and I in Newark, NJ, 1940	6
Learning how to drive a tractor in Tarkio, Missouri, in 1941	7
Dressed in western garb on the lecture circuit in the Cornbelt in 1941	8
As a graduate student at the University of Wisconsin, 1944	10
Kurt Mislow's retirement party at Princeton in June 1988	11
Machete duel with Gilbert Stork in Mexico in 1954	12
The mescaline party in my yard in Birmingham, Michigan, in 1954	15
At the Tuborg Brewery in Copenhagen in 1962	16
With my boy scout troop in the Vienna Woods in 1936	19
My research advisor, A. L. Wilds, in 1944	20
Pedro Lehmann, Russell Marker, Roger Adams, and I in 1969	24
My research laboratory at Syntex in Mexico City in 1950	26
Louis and Mary Fieser in the late 1940s	32
Press conference at Syntex in Mexico City, 1951	34
Gilbert Stork in Paris in 1954	37
David Lightner, Keith Brown, and I at Stanford in 1961	38
With R. B. Woodward and I. V. Torgov at the Prague IUPAC Conference in 1962	39
National Medal of Science Ceremony at the White House in 1973	41
At a Columbia University commencement in 1974 with Frances Hoffman	42
At the Gordon Conference on Steroids and Related Natural Products, summer of 1953	45

Arthur Birch lecturing in the United States in 1954	47
Induction into the National Inventors Hall of Fame	51
With Alejandro Zaffaroni, 1959	53
With Huang Liang, my first Chinese postdoctoral fellow, in Beijing	54
With my first Ph.D. student, Carl Lenk, in 1952	55
Leopold Ruzicka in Zurich on February 14, 1951	57
A Djerassi group at Wayne University in 1955	58
W. Moffitt and R. B. Woodward in Woodward's office in 1958	60
A Djerassi group at Wayne State University in 1957	61
In Mexico City in 1951 at Syntex	62
At Wayne State University in 1957	64
At a 1990 ski excursion in the Sierras	66
Press conference in my Stanford office in 1961	68
Discussing a joint publication at Stanford in the early 1960s	71
With Günther Snatzke at an IUPAC meeting in Riga in 1970	75
At the IUPAC Natural Products Meeting in Sydney, Australia, 1960	78
Australian chemist K. E. Murray demonstrating instrumentation	79
At the 1964 IUPAC meeting in Kyoto, Japan	85
Ken Rinehart, Al Burlingame, and Fred McLafferty in April 1965	86
Opening of the Syntex Institute of Molecular Biology, circa 1962	103
The visit of King Carl XVI Gustaf of Sweden	105
With Ed Feigenbaum at his 50th birthday party in 1986	106
An annual cross-country skiing trip with my research group	108
My son, Dale, and I about to go diving in Pacific Grove, CA	122
With the remnants of a lobster aboard a Soviet research ship in 1989	124
One of the skiing trips with my students in 1988	137
Aboard Ben Tursch's boat in Papua New Guinea	139
With members of the Taiwanese feminist organization called Awakening	144
One of my many trips to Brazil	146
With Georgi Arbatov at the Pugwash Conference in the USSR in 1969	148

With Thomas Odhiambo at one of our many ICIPE meetings in Nairobi 149

With Koji Nakanishi and Jerrold Meinwald preparing to go to Tanzania 150

Howard Ringold and I with Syntex consultants in the mid-1960s 151

At my 60th birthday celebration 152

Derek Barton and I with Finn Sandberg and two Korean colleagues 153

With Robert Maxwell in Oxford on the occasion of his 60th birthday 154

With Koji Nakanishi in Sendai in 1964 155

Koji Nakanishi and I at a symposium in 1990 156

My first and only triumph over Koji Nakanishi's magic in 1990 157

With Andre Dreiding at the Picasso Museum in Lucerne 158

Four generations of Djerassi males 159

Reunion with old friends in October 1988 160

My first literary book-signing of my novel, *Cantor's Dilemma* 161

Trekking in Bhutan near the base camp to Chomo Lhari 162

With Valery Galkin near Vladivostok in September 1989 163

With Diane Middlebrook at our wedding in June 1985 164

My son Dale and my daughter Pamela 165

A dance performance at the Djerassi Foundation's Artist Colony 166

With Victoria Morgan and Eda Holmes after the dance 167

With poet Robert Lowell at a degree ceremony at Kenyon College 168

Opera star Tatiana Troyanos autographing my short story 168

My first wife, Virginia, whom I married before my 20th birthday 169

My second wife, Norma, the mother of my two children 169

Picture taken in 1989, after I wrote my last autobiographical memoir 170

Preface

"HOW DID YOU GET THE IDEA—and the good fortune—to convince 22 world-famous chemists to write their autobiographies?" This question has been asked of me, in these or similar words, frequently over the past several years. I hope to explain in this preface how the project came about, how the contributors were chosen, what the editorial ground rules were, what was the editorial context in which these scientists wrote their stories, and the answers to related issues. Furthermore, several authors specifically requested that the project's boundary conditions be known.

As I was preparing an article[1] for *Chemical Reviews* on the Curtin–Hammett principle, I became interested in the people who did the work and the human side of the scientific developments. I am a chemist, and I also have a deep appreciation of history, especially in the sense of individual accomplishments. Readers' responses to the historical section of that review encouraged me to take an active interest in the history of chemistry. The concept for Profiles, Pathways, and Dreams resulted from that interest.

My goal for Profiles was to document the development of modern organic chemistry by having individual chemists discuss their roles in this development. Authors were not chosen to represent my choice of the world's "best" organic chemists, as one might choose the "baseball all-star team of the century". Such an attempt would be foolish: Even the selection committees for the Nobel prizes do not make their decisions on such a premise.

The selection criteria were numerous. Each individual had to have made seminal contributions to organic chemistry over a multidecade career. (The average age of the authors is over 70!) Profiles would represent scientists born and professionally productive in different countries. (Chemistry in 13 countries is detailed.) Taken together, these individuals were to have conducted research in nearly all subspecialties of organic chemistry. Invitations to contribute were based on solicited advice and on recommendations of chemists from five continents, including nearly all of the contributors. The final assemblage was selected entirely and exclusively by me. Not all who were invited chose to participate, and not all who should have been invited could be asked.

A very detailed four-page document was sent to the contributors, in which they were informed that the objectives of the series were

1. to delineate the overall scientific development of organic chemistry during the past 30–40 years, a period during which this field has dramatically changed and matured;

2. to describe the development of specific areas of organic chemistry; to highlight the crucial discoveries and to examine the impact they have had on the continuing development in the field;

3. to focus attention on the research of some of the seminal contributors to organic chemistry; to indicate how their research programs progressed over a 20–40-year period; and

4. to provide a documented source for individuals interested in the hows and whys of the development of modern organic chemistry.

One noted scientist explained his refusal to contribute a volume by saying, in part, that "it is extraordinarily difficult to write in good taste about oneself. Only if one can manage a humorous and light touch does it come off well. Naturally, I would like to place my work in what I consider its true scientific perspective, but . . ."

Each autobiography reflects the author's science, his lifestyle, and the style of his research. Naturally, the volumes are not uniform, although each author attempted to follow the guidelines. "To write in good taste" was not an objective of the series. On the contrary, the authors were specifically requested not to write a review article of their field, but to detail their own research accomplishments. To the extent that this instruction was followed and the result is not "in good taste", then these are criticisms that I, as editor, must bear, not the writer.

As in any project, I have a few regrets. It is truly sad that Egbert Havinga, who wrote one volume, and David Ginsburg, who translated another, died during the development of this project. There have been many rewards, some of which are documented in my personal account of this project, entitled "Extracting the Essence: Adventures of an Editor" published in CHEMTECH.[2]

Acknowledgments

I join the entire chemical community in offering each author unbounded thanks. I thank their families and their secretaries for their contributions. Furthermore, I thank numerous chemists for reading and reviewing the chapters, for lending photographs, for sharing information, and for providing each of the authors and me the encouragement to proceed in a project that was far more costly in time and energy than any of us had anticipated.

I thank my employer, Philip Morris USA, and J. Charles, R. N. Ferguson, K. Houghton, and W. F. Kuhn, for without their support, Profiles, Pathways, and Dreams could not have been. I thank ACS Books, and in particular, Robin Giroux (acquisitions editor), Karen Schools Colson (production manager), Janet Dodd (senior editor), Joan Comstock (department head), and their staff for their hard work, dedication, and support. Each reader no doubt joins me in thanking 24 corporations and Herchel Smith for financial support for the project.

I thank my wife Suzanne, for she assisted Profiles in both practical and emotional ways. I thank my children Jonathan and Brooke for their patient support and understanding; remarkably, I have been working on Profiles for more than half of their lives—probably the only half that they can remember! My family hardly knows a husband or father who doesn't live the life of an editor. Finally, I again thank all those mentioned and especially my family, friends, colleagues, and the 22 authors for allowing me to share this experience with them.

JEFFREY I. SEEMAN
Philip Morris Research Center
Richmond, VA 23234

February 15, 1990

[1] Seeman, J. I. *Chem. Rev.* **1983**, *83*, 83–134.
[2] Seeman, J. I. *CHEMTECH* **1990**, *20*(2), 86–90.

Editor's Note

CARL DJERASSI played an instigative role in the ACS's decision to expand the Profiles series from the one-volume book that was originally planned. Djerassi's chapter was one of the first completed and, at the time, the longest. He had been pressured to cut it. He responded, "The good news is that I have now finished my chapter.... The bad news is that my chapter is way beyond your page limit and that I see no acceptable way to reduce it to any significant extent. This leaves us with three options: (a) You ... convince the ACS to issue a multi-volume book ... (b) We eliminate the epilogue and thereby save nearly 17 pages ... (c) I withdraw my contribution from your book. I shall hold no hard feelings whatsoever if you make this choice, because the excessive length is solely my fault and it would be totally unfair to treat me differently from your other authors." Djerassi demanded a rapid decision, and it was immediately forthcoming: The ACS decided that "the content of Profiles is more important than any predetermined book size."

Djerassi pushed the page limitation issue, but he was willing, without rancor, to pull his manuscript. This was an impressive sequence of events, for it demonstrated his ability to see beyond the current status and his determination to have issues answered or defined. If he is not in a position to make a decision unilaterally, he will then force those who are to do just that. He will commit, but only if he is satisfied that the result will live up to his high standards.

On May 27, 1988, following a year of additional writing and modifications, Djerassi mailed the final version of his manuscript, together with a cover letter. "The enclosed manuscript incorporates virtually all requests that you made.... I hope that you are impressed.... While I will be happy to read any letters from you that acknowledge receipt of this manuscript, or any complimentary remarks, *do not even think* of writing me another letter requesting any more changes..."

Some months later, Koji Nakanishi recommended Djerassi as the independent reader for his chapter. Nakanishi's suggestion was excellent because the Djerassi–Nakanishi ties extend back more than 30 years. Nakanishi had translated Djerassi's ORD book into Japanese;

Djerassi "arranged for the English publication of [Nakanishi's] first book on infrared spectroscopy." When Djerassi was president of Syntex Research and, later, CEO of Zoecon, Nakanishi was a consultant for both companies.

Critically reading these chapters is a demanding and time-intensive task. Because Djerassi is clearly an overcommitted individual, and because he was *so* specific regarding any additional requests from me, I was reluctant to request that he expend additional time on Profiles. Thus, I soon reversed my decision that he review Koji's manuscript. Surprised was I to receive Djerassi's telephone call. "Where is Koji's manuscript?" Djerassi demanded. "Koji asked me to read it; I am waiting for it; have you not sent it?"

Shortly thereafter, Djerassi called again and enthusiastically described his fascination with Nakanishi's manuscript: "I could hardly put it down. I certainly learned a great deal about [him] that I did not know." This statement was remarkable, given their long-standing friendship. However, it was not to be unique: Other readers of early drafts of these autobiographies responded similarly, learning new facets and facts about their old friends.

Djerassi was animated about Profiles, impressed that other contributors had spent as much time as he in penning their volumes. This was a crack in the facade, for as good and complete as Djerassi's manuscript was, his recently published short pieces (both fiction and nonfiction) suggested that there was an aspect of the man not yet reflected in his chapter.

"When I first got your request of December 1, 1988, to write a coda to my chapter, I was certain that I would not accede. As far as I was concerned, that chapter was finished and done with," Djerassi wrote on December 27th. Enclosed with that letter was the coda to his chapter. Djerassi was somewhat apologetic. "I did produce a coda of some sort, although it may not be at all what you had in mind." In fact, it was more, infinitely more than anticipated. It was so powerful that I could hardly move from my chair. The coda is a powerful reflection of the Djerassi philosophy.

Djerassi's efforts to enhance his Profiles volume did not stop with his coda. He subsequently added more textual inserts and more photographs and provided more robust photo captions. He closely monitored the progress of the series and continually provided suggestions for its improvement. Believing that controversy and "spice" would encourage sales, at the last minute, Djerassi changed the title of his volume from *A View Through Steroid Glasses* to *Steroids Made It Possible*.

Perhaps my most amusing interaction with Djerassi occurred just as the final deadline arrived. I had to check with him regarding the

Carl Djerassi, William S. Johnson, and Gilbert Stork chat during a symposium at Stanford.

very last corrections, and he returned my call immediately. When the last matters had been resolved, I asked him how he felt about his fused knee. Djerassi immediately responded, in his typically striking and extraordinary fashion, with a revealing discourse on the positive aspects of his handicap.

"Would you add that to your chapter?" I asked.

"SEEMAN!" he shouted emphatically, "I am calling you from my car, I am on the way to the airport, I have done more for you than I have done for any other editor . . . "

"Okay, okay," I responded, "just call up your secretary, repeat what you have just told me, and have her fax it to me." Within the hour, I received a fax that began "The following item was dictated by Dr. Djerassi while driving to the airport, and as a result was not checked by him for inaccuracies." The caption for the photograph on page 66 followed.

We simply do no have enough information here to resolve, to evaluate, to grasp the totality of Carl Djerassi. The content of his many hundreds of scientific publications is only hinted at in his book. The impact of his work on steroid chemistry, his development of spectroscopic methods, his participation in the modern era of the pharmaceutical age, and his contributions to scientific publishing cannot be easily encompassed simultaneously. One gains further insight by reading, or perhaps more accurately, by absorbing and assimilating his fiction and his "not-so-fiction". But there is more to this man, far more.

Djerassi's enormously strong ego is moderated by his sense of humor: He ends his scientific autobiography by referring to his scientific literature as "clutter[ing] up *Chemical Abstracts* or *Current Contents.*" Has there been so much of Carl Djerassi that even he cannot manage the accounting?

June 7, 1990

Steroids Made It Possible

Carl Djerassi

Diary Entry (11 August 1983)

We sit in Copenhagen,
Chemists from a dozen countries.
The talk is heavy; the words are long:

Male contraception,
Cures for cancer,
Morphine substitutes,
Drugs from the sea,
Medicines for the year 2000.

We've mouthed these words for many years,
Formulae hiding the chemists.
Who are these colleagues, students, strangers?
What do they do besides chemistry?

If this were the Holiday Inn,
Not the Royal Danish Academy,
Would I guess who they are?

A convention of grocers? Too serious.
Car salesmen? Too little polyester.
Bankers? Lawyers? No vests.
Clergymen? Wrong collars.
Poets? Nobody smokes.

How did they come to chemistry?
What do they do besides chemistry?
What do I do besides what I do
Besides chemistry?

C. Djerassi, *SANDS 1987*, pp 10–11

H OW DID I GET INTO CHEMISTRY? I didn't have any childhood chemistry sets; I never blew up our basement; I never had chemistry in high school (in fact, I never graduated). Until age 14, I attended a *Realgymnasium* in Vienna. After the Hitler *Anschluss* I left for my father's home in Sofia, Bulgaria, and entered a private school, The American College, where most subjects were taught in English. Both my Bulgarian father and Viennese mother, by then divorced, were physicians who had met in medical school in Vienna; all of us had assumed that I would go into medicine.

My mother and I arrived penniless in the United States in December 1939, shortly after my 16th birthday. We landed in New York City aboard the liner *Rex*, soon to be sunk in the war. Within a month, I entered the now-defunct Newark Junior College in Newark, New Jersey. The choice was pure serendipity. One of my American teachers in Sofia had recommended that I consult one of his friends teaching at New York University about further schooling. Entrance into New York University that late in the academic year proved impossible, but he arranged for Newark Junior College to accept me at the start of the January semester. While my mother (who had no license to practice in the United States) found employment in upstate New York as a doctor's assistant, I was placed through a refugee aid organization into the Newark home of an extraordinarily generous and kind family. My quick and relatively painless adjustment to America was helped along by Frank Meier, an inorganic chemist working at Engelhardt Industries; his wife, Clara, a local high school teacher; and their two high-school-aged sons, August (now a professor of history at Kent State University) and Paul (now a professor of statistics at the University of Chicago).

My father, Samuel Djerassi, and I walking past the headquarters of the Bulgarian Academy of Sciences in 1939 in Sofia.

I compressed my college career into five semesters: two in Newark; one semester at Tarkio College in Missouri, where I had received a room, board, and tuition scholarship; and two semesters plus a summer at Kenyon College in Ohio.

My mother, Alice Djerassi, and I shortly before our arrival in New York City in December 1939.

Early in 1940, I expressed my need of a scholarship to continue my schooling by writing to Eleanor Roosevelt, whom I felt was the individual most likely to make things happen with a mere wave of her wand. A brief reply arrived from the Institute for International Education, to which she had forwarded my request, followed sometime later by a postcard from an officer of the institute. "I have some good news for you. You have been awarded a scholarship for the next semester at Tarkio College in Tarkio, Missouri."

Soon, I announced to my mother that I was heading west to a town I couldn't find on any map then at my disposal. Even the Greyhound ticket office in New York had some difficulty before discovering that the northwestern corner of Missouri was my ultimate destination. St. Joseph was the city where Greyhound deposited me after changes in Pittsburgh and Kansas City, and where I transferred to a local bus that eventually passed the sign "Welcome to Tarkio, Queen of the Cornbelt." That was the town where I launched my career as a public speaker at the age of seventeen.

Within hours of my registration at Tarkio College—a Presbyterian school where a God-fearing student body of

140 was guided by a faculty of 20 (including the business manager, a house mother, and the superintendent of grounds)—I learned a piece of historical lore which seemed a superb omen to a budding chemistry major like myself: Tarkio's most illustrious alumnus was none other than Wallace Carothers, the inventor of nylon. No one mentioned that he'd committed suicide just a few years after making his discovery.

Paul Meier (now a professor of statistics at the University of Chicago) and I, shortly after my arrival in the United States (Newark, NJ, 1940).

At the end of my first week, the program chairman of the Tarkio Rotary Club informed me that I was scheduled to address the local members who had contributed to my scholarship on "The Current Situation in Europe." I was too nervous and too innocent of public speaking even to imagine the many pitfalls a teenaged, city-bred kid might face in front of an audience of middle-aged farmers and businessmen. However, I was not too innocent to plagiarize ruthlessly John Gunther's *Inside Europe*. His reference to the Balkans—"Must every little language have a country all its own?"—still sticks in my mind. My peculiar accent gave the speech an air of authenticity that no one—even one who knew the facts—could possibly resist.

At the end of my Rotary debut, the minister of the local Presbyterian church congratulated me on my performance and proposed that I give a similar talk to his congregation after his Sunday sermon. In order not to repeat myself, I rushed back to my source—*Inside Europe* was not only up to date but also full of entertaining tidbits—and shamelessly borrowed new material which I dressed up in my Viennese–Bulgarian–British accent. This second talk propelled me almost overnight from amateur to professional status. In the minister's office that Sunday, with an apologetic reference to the size of the offering, my pleased host

My first experience with a motor vehicle was learning how to drive a tractor in Tarkio, Missouri in 1941.

shoved the contents of the collection plate—nickels, dimes, and occasional quarters—in my direction. It was my first lecture honorarium.

The ministerial grapevine apparently crossed denominational boundaries. From that Sunday on I received almost weekly invitations to address various church groups in northwestern Missouri and southwestern Iowa about "The European Situation." My plagiarism of Gunther became more sophisticated. I mixed his journalistic wisecracks, such as his definitions of Balkan peace ("a period of cheating between two periods of fighting") or Balkan revolutions ("abrupt changes in the form of misgovernment"), with references to my personal reminiscences from life in prewar Bulgaria. I boasted about the amazing quality of the Sofia

Dressed in western garb while on the lecture circuit in the Cornbelt in 1941.

Opera and the fact that Bulgaria was the source of most of the world's attar of roses.

As the weeks passed, the collection plates became the chief source of my pocket money. Instead of waiting on tables or performing other plebeian part-time work, I pontificated to ladies' auxiliaries, church congregations, and even an eighth-grade commencement ceremony with the self-assurance of a so-far-undetected plagiarist.

From "Dear Mrs. Roosevelt"[320]

Virtually by default, I started out as a premed major but what converted me into a chemist were a superb chemistry teacher at Newark Junior College, Nathan Washton (now professor emeritus at Queens College); the equally superb two-man Kenyon College Chemistry Department, consisting of Walter H. Coolidge and Bayes M. Norton, in classes ranging from two to four students; and my lack of financial resources for medical school. My undergraduate research with Norton at Kenyon College had been in physical chemistry, focused on the quantum yield in the photolysis of ethyl iodide, but the seeds for a career in organic chemistry were probably sown indirectly by my European childhood among doctor parents and my reading of Paul de Kruif's *Microbe Hunters*. The final push was my first year of medicinal chemistry research at CIBA before going on to graduate school.

Without a shred of embarrassment I acknowledge my intellectually polygamous nature. It is not only chemistry that stimulates and excites me. So does art, which I collect; music, which I used to play; and *belles lettres*, which I read and write. Starting at age 18, I have led a thoroughly polygamous life, even within my own scientific discipline of organic chemistry. Thus, over a period of four and a half decades, I have been active in the development of new medicinal agents (antihistamines, oral contraceptives, and topical corticosteroids) that are still being used by millions; the isolation, structure elucidation, and partial synthesis of several hundred natural products (steroids, lipids, terpenoids, alkaloids, and antibiotics); certain aspects of their biological function and biosynthesis; and, perhaps most importantly, the development of physical methods (such as optical rotatory dispersion, optical and magnetic circular dichroism, mass spectrometry, and computer artificial intelligence techniques) for the solution of stereochemical and structural problems.

Nevertheless, even in this harem of chemical infatuations, one consistent, unswerving love never left me: steroids. Therefore, I shall describe my organic chemical research primarily through the eyes of a chemist who saw steroids as both paint and canvas, and physical methods as the brush.

As a graduate student in front of the chemistry building at the University of Wisconsin, 1944.

In 1942, when I embarked on my research career, the only physical method used by organic chemists was ultraviolet spectroscopy, involving laborious point-by-point measurements. The University of Wisconsin Chemistry Department, where I did my graduate work, did not even boast of a Beckman DU ultraviolet spectrophotometer. I had to walk partway across campus to obtain access to one. The only separation method, other than fractional distillation and recrystallization, was column chromatography, and I may well have been the first chemistry graduate student at Wisconsin to have used it extensively. Gas chromatography, thin-layer chromatography (TLC), and high-performance liquid chromatography (HPLC) (to name just three techniques) were still unknown. I ran microanalyses myself. (Thirty years later, I argued unsuccessfully with a foreign journal editor that microanalyses were largely useless—other than to demonstrate the presence of dirt or filter paper in the sample—in the face of easily accessible high-resolution mass spectral data.) Using currently available separation techniques and sophisticated physical methods, today's natural products chemist can frequently establish the complete structure of a complex molecule with only micrograms of material. We have now reached the stage where often we have insufficient material for a retention sample; where crystallization is not worth attempting; where determination of a melting point may be a prohibitive waste of material; and yet, where we have learned more about the structure of that molecule than we did years ago with grams of substance.

As a research scientist, I have published a great deal; too much, some of my peers have said. The most amusing, tongue-in-cheek description came in 1963 from Kurt Mislow[1]:

> The vast outpouring of publications by Professor Djerassi and his cohorts marks him as one of the most prolific scientific writers of our day. . . . a plot of N, the papers published by Professor Djerassi in a given year, against T, the year (starting with 1945, $T = 0$) gives a good straight-line relationship. This line follows the equation $N = 2.413T + 1.690$. . . . Assuming that the inevitable inflection point on

Kurt Mislow's retirement party at Princeton in June 1988. Standing: Jerry Berson, Al Moscowitz, Jeremy Knowles, Carl Djerassi, and Paul von Ragué Schleyer. Front row: Duilio Arigoni, Kurt Mislow, and Al Meyers.

the logistic growth curve is still some 10 years away, this equation predicts (a) a total of about 444 papers by the end of this year, (b) the average production of one paper per week or more every year beginning in 1966, and (c) the winning of the all-time productivity world championship 10 years from now, in 1973. In that year Professor Djerassi should surpass the record of 995 items held by . . .

"Thank God," many chemists will say gratefully, "Mislow's prediction turned out to be off by 8 years." Still, it should not be totally surprising that in 45 years I became the coauthor of over 1100 papers. In my academic career, I usually had around 20 collaborators annually,

Staging a machete duel with Gilbert Stork (left) while collecting giant cacti in Mexico in 1954 for triterpene studies at Wayne University (later called Wayne State University).

which meant that cumulatively well over 300 pre- and postdoctoral colleagues from 51 countries have passed through my laboratory. In addition, for many years—some of them concurrent with my academic life—I led a productive industrial research program, which also resulted in much publishable work. I have always felt firmly that if I persuade another individual to collaborate with me on a specific scientific project, which is successfully completed, then in fairness to that person's professional advancement, the work should be written up. Indeed, it has often been said (correctly so, in my opinion) that scientific work is not complete until it has passed the rigor of peer review and been published. A scientist should contribute to the pool of published knowledge, not just draw from it.

With very few exceptions, I myself have written the initial drafts of the first 500 or so papers. Only in the 1960s did I conclude that teaching my younger collaborators how to prepare a publishable scientific paper was also one of my pedagogic functions. Since that time, most of the first drafts were prepared by my many colleagues, which is another explanation for the relatively large number of publications that have emanated from my laboratory.

Clearly, the following account can cover only a fraction of that work. Selecting certain examples does not imply that I necessarily value these projects more than others, although I am certainly picking some that I consider to be high points in my chemical life. Rather, I shall document that if one common thread can be discerned throughout the fabric of my scientific contributions, it was and is my use of steroids for biological, synthetic, physical, and mechanistic purposes. As a result, substantial areas of natural products research will be skipped altogether or touched upon only briefly.

I shall mention none of our work—conducted over nearly a decade—on macrolide antibiotics, which included the first complete structure elucidation[2] of a member of that important class, methymycin (1). Or our studies of pentacyclic triterpenes from giant cacti,[3] which

1

was initially prompted by my exposure to these plants while performing steroid research in Mexico. Nor will it be possible to cover our alkaloid work, recorded in nearly 70 papers, starting from our first Mexican cactus alkaloid, pilocereine (2),[4] and ending in 1974 with alkaloids[5]

2

derived from a plant growing on my own California ranch. These structure elucidations, eliminated from the present account, have meant a lot to me and at the same time have produced many challenging synthetic and biosynthetic targets for others. I have reviewed this work elsewhere[6] under the title "Natural Products Chemistry 1950 to 1980—A Personal View." It is with some relish that I recall the flood of reprint requests prompted by the following footnote on the title page: "Selected personal statements by the author were removed by the editor without Professor Djerassi's consent. An uncensored version of this paper can be obtained by writing to Professor C. Djerassi . . ."

One personal episode from my early alkaloid days may be worth telling, because it will show something of my nonchemical side during those years. As one of the very few persons in the world then working on new cactus alkaloids, I felt that I, an individual previously uncontaminated by exogenous behavioral modifiers, should experience personally the biological effects of that most famous of all cactus alkaloids: mescaline. At one of my research seminars, I announced that the thesis (mescaline ought to become the opium of the intelligentsia) of Aldous Huxley's *The Doors of Perception* would be tested experimentally the following Sunday afternoon at a picnic in my backyard. Were there any volunteers? To my surprise, only two offered to join me: one was a postdoctorate research fellow, Sandy Figdor, now at Pfizer; the other was a graduate student, Marvin Gorman, now vice president of Bristol–Myers.

The menu for the Sunday picnic was pizza for the students and spouses, pizza-*cum*-mescaline sulfate for the three experimentalists. At least that was our plan. In fact, the mescaline sulfate was so extremely bitter that we took it with orange juice. I felt sufficiently squeamish to skip the pizza. Instead, I lay down in the grass to wait for my experiences. Relaxing was not easy with at least 20 pizza eaters looking out of the corner of their eyes at me and my two fellow guinea pigs. The previous day, I had re-read the section on mescaline in the 1953 edition of *The Alkaloids: Chemistry and Physiology*. That otherwise dry compendium raved about the effects of mescaline in words rarely found in chemical treatises: ". . . ordinary objects appear to be marvelous . . . sounds and music are 'seen' in color . . . color symphonies and new, unknown colors of unimaginable beauty and brilliancy are perceived."

I stared at the flowers in our garden, periodically closing my eyes and trying to retain the image of the flower I had last possessed visually. I had no luck. The flowers did not look any different, with eyes closed or open, nor were the colors any more intense than in my previous, drugless life. I turned on the record player on the porch. Mozart's C-Minor Piano Concerto was waiting for the needle. While Mozart

The infamous mescaline party in my backyard in Birmingham, Michigan, in 1954. My graduate students and their wives are waiting for the mescaline transformation of Carl Djerassi.

resounded in the garden, I felt only the nausea that crested in my stomach every time one of my pepperoni-smelling students asked, "What's it like?"

Somewhere between 2 and 3 hours passed, during which time it had become clear to this colloid of young chemists, who had been throwing me sidelong glances or dropping pointedly apposite remarks concerning color or sound, that I was turning into a taciturn host. As the first few students started to depart, I rose to accompany them to the gate and the first effect hit me. With each step I seemed to be floating, weightlessly, toward the street. Pretending uncharacteristic chumminess, I put my arms around the shoulders of two students as they walked to their car. They did not know I needed their ballast to prevent me from sailing up into the sky.

Eventually, only six of us were left: three mescalinized men, who for some reason had not talked to each other until that moment,

I had never touched beer until I was almost 50 years old. This picture was taken in 1962 at the Tuborg Brewery in Copenhagen at a lecture before the Danish Chemical Society. Fortunately, Tuborg produced an alcohol-free beer, with which I proposed a toast to the Danish chemists. Photo courtesy of Associated Press.

and three sober and somewhat worried wives. As we started to compare notes, we realized that embarrassment had kept us apart. Each had been afraid to learn from the others that he was the sole mescaline failure. The resulting cacophony of laughter was not solely an expression of relief, but also a manifestation of our delayed response to mescaline.

I still blush, however, at a moment I would normally have considered a sad display of pedagogic boorishness. With tears running down my face and my mouth full of food, I regaled my audience with my vision of a forthcoming Ph.D. examination on a massive dissertation I had just read. My companions all knew the candidate and knew he deserved to pass. What, I croaked triumphantly to my two doubled-up companions, while the three wives stared stonily, would he do if, upon completion of the oral examination, I rose with a sardonic smile on my own face and tore his telephone-book-sized thesis in two parts as he reached out to shake my hand? For the only time in our marriage, my wife restrained me physically as I rose to get our telephone directory in order to demonstrate my calling as a circus strong man.

Even as I, the lifelong teetotaler, behaved like a drunk, another wave of the mescaline experience descended upon me. The "real" me seemed to be sitting in one corner of the room observing, coolly, the spectacle of the other me acting without inhibition. Aldous Huxley quoted this line in William Blake's *The Marriage of Heaven and Hell*: "If the doors of perception were cleansed every thing would appear to man as it is, infinite." I reached a different conclusion: it takes more than mescaline, or indeed chemistry, to change a man's persona.

The Partial Synthesis of Steroids

Partial Aromatization of Androgens to Estrogens

In 1962, I persuaded 16 graduate students at Stanford University to prepare under my editorship a book entitled *Steroid Reactions*.[7] In my introduction, I cited the following two reasons for the value of such a book.

> First, there is practically no newly discovered synthetic organic reaction which is not applied almost immediately in the steroid field and where, as a consequence of the requirements of great selectivity in polyfunctional situations, reaction conditions are modified. These conditions invariably offer greater scope and are frequently "gentler" in nature.

Second, because of the peculiarities of the steroid field—the large number of synthetic transformations beginning with only a handful of different steroidal starting materials—many organic reactions and organic protecting groups have been developed specifically in this field or have found their widest and most diverse applications among steroids. Frequently, they just wait to be discovered by the non-steroid specialist.

In many ways, these sentences describe the leitmotif of my first decade of chemical research, even though most of the projects were prompted by a search for clinically useful molecules.

Because of an early skiing accident in Europe, which eventually led to a knee fusion, I was ineligible for military service and had the luxury of a college education while most of my contemporaries ended up in the armed forces. I had graduated in 1942 from Kenyon College in Ohio, not yet 19, and had started working as a junior chemist at CIBA Pharmaceutical Products, Inc., in Summit, New Jersey. As luck would have it, I worked for 1 year in the laboratory of Charles Huttrer, another Hitler refugee from Vienna, who, though 20 or more years older than I, treated me like an equal. Together we discovered one of the first antihistamines—Pyribenzamine (tripelennamine) (3)[8]—by synthesizing a series of ethylenediamines, which were screened pharmacologically by another European refugee, Rudolf L. Mayer, who had

$$\underset{N}{\bigcirc}\!\!-\!\!\underset{|}{N}\!\!-\!\!CH_2CH_2N(CH_3)_2$$
$$\phantom{\underset{N}{\bigcirc}\!\!-\!\!N\!\!-\!\!}CH_2C_6H_5$$

3

brought the concept of histamine's role in allergic reactions to CIBA management's attention. The rapidity with which this substance became the drug of choice for hundreds of thousands, if not millions, of allergy sufferers spoiled me for a long time; I became an inveterate optimist as far as scientific success was concerned.

The Swiss parent of CIBA was one of the original powerhouses in steroid chemistry and medicine, the work in Basel having been directed first by K. Miescher and later by A. Wettstein. Thus, even though my first year at CIBA was dedicated exclusively to antihistamines, I was exposed to steroids through contacts with some of my older laboratory colleagues. I also read the first edition of Fieser's *Natural Products Related to Phenanthrene*.[9] This highly readable book, more than any other factor, got me permanently hooked on steroids.

I learned to ski in Austria. Here I posed (second from left) with my boy scout troop in the Vienna Woods in 1936, two years before my skiing accident.

A couple of graduate courses at night—first at New York University and then at the Brooklyn Polytechnic Institute—convinced me that part-time graduate study, which involved commuting to New York City after a full day at CIBA and gulping down a dinner snack on the Lackawanna Railroad, was no way to get a Ph.D., at least not for someone with my sense of urgency. After 1 year at CIBA, I decided to accept a Wisconsin Alumni Research Foundation fellowship at the University of Wisconsin for work toward a Ph.D. degree.

In 1943, the University of Wisconsin Chemistry Department had two young assistant professors, A. L. Wilds and W. S. Johnson, who were about to undertake ambitious projects dealing with the total synthesis of steroids. At that time only one steroid hormone had been synthesized totally—equilenin (18)—and Wilds had been one of the members of the famous Bachmann, Cole, and Wilds team responsible for that feat at the University of Michigan.[10] I chose Wilds as my advisor and selected for my Ph.D. thesis project a compromise between Wilds's interests in total synthesis and my budding one (based on the Fieser book[9] and CIBA lore) in chemical transformations of intact steroids: the partial aromatization of androgenic steroids to the estrogens. In retrospect, this choice had amazingly long-term consequences on me.

In specific terms, the problem could be defined as the transformation of the male sex hormone testosterone (5) into the female sex hormone estradiol (6a). In the early 1940s, this problem was of major practical significance: testosterone (5) was available industrially by partial

My research advisor, A. L. Wilds, at that time (1944) assistant professor at the University of Wisconsin.

synthesis from cholesterol (4), whereas estradiol (6a) still had to be extracted laboriously from pregnant mare's urine. At first sight, my Ph.D. thesis topic appeared to be pertinent solely to the steroid field. However, if the question is posed in more general terms—how can one ring in a polycyclic molecule be selectively aromatized?—then it becomes a problem in general organic chemistry. If, in addition, it is noted that such selective aromatization is blocked by two quaternary centers (C-10 and C-13 in 5), then the problem becomes virtually unprecedented, especially if good yields are required.

Just prior to the outbreak of World War II, H. H. Inhoffen in Germany[11,12] had presented one possible solution to this problem, based on a precedent from sesquiterpene chemistry: the acid-catalyzed conversion of santonin (10) to desmotroposantonin (11). By conversion of cholesterol (4), or the corresponding androstene derivative, into the saturated ring-A ketone 7a, followed by dibromination (7b) and dehydrobromination, a steroidal 1,4-dien-3-one (8) was obtained. Exposure to

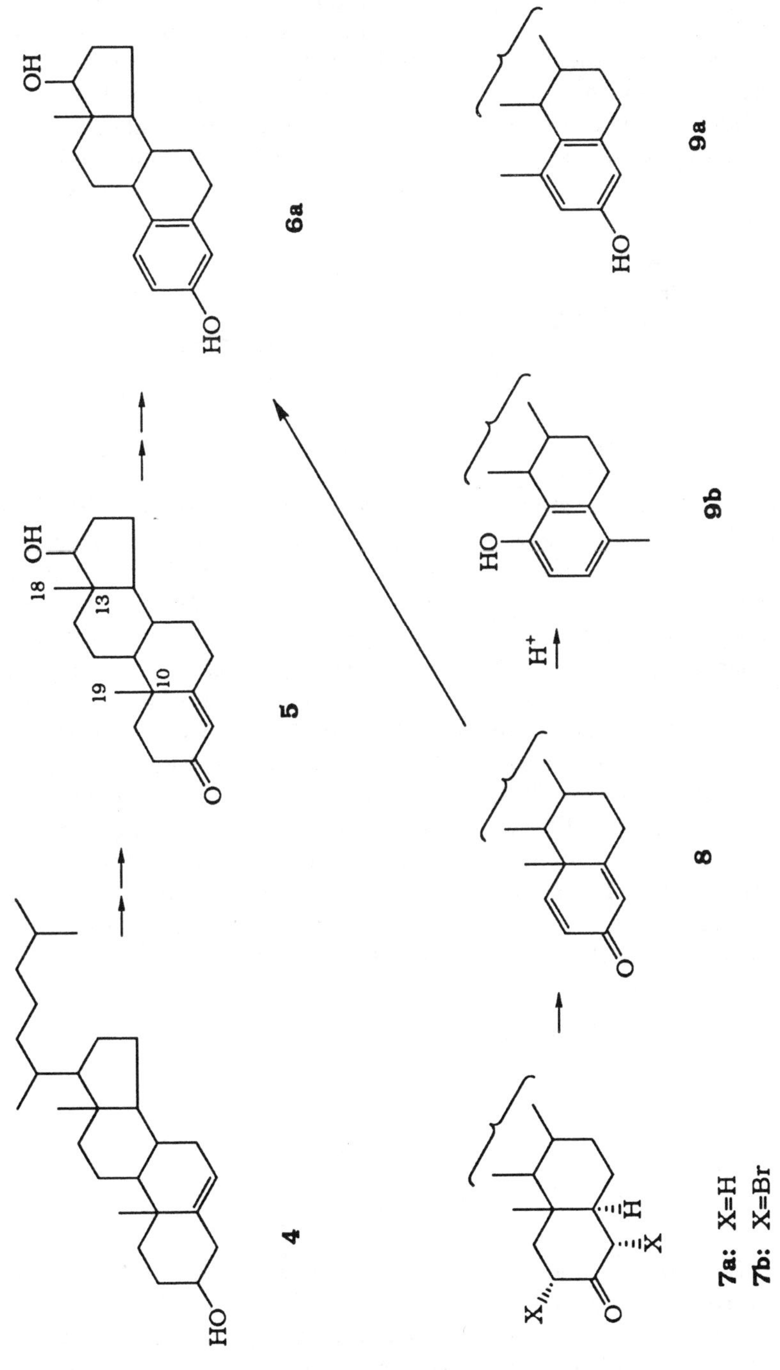

OH
HO
6a
18
OH
13
19
10
O
5
HO
4
9a
HO
HO
9b
H+
O
8
X
O
H
X
7a: X=H
7b: X=Br

acid resulted in methyl migration and partial aromatization to yield a phenol, which was assumed by Inhoffen to possess the 1-methyl-3-hydroxy aromatic nucleus 9a, on the basis of analogy to the well-known desmotroposantonin (11). Even more importantly, Inhoffen found[11-13] that when the dienone 8 in the androstane series was vaporized in the presence of tetralin, selective aromatization of ring A proceeded with elimination of the angular methyl group to produce estradiol (6a).

As a graduate student, I was able to confirm Inhoffen's first partial synthesis of an estrogen, except for the reported 60% yield in the final aromatization step, for which no experimental details had been published.[13] In order to gain more insight into such partial aromatizations, my thesis advisor Wilds suggested that I study a model other than santonin (10)—specifically the chrysene 12. Not only did I succeed[14] in synthesizing the polycyclic dienone 12 and in proving the structure of the aromatic phenol 13, produced by acid catalysis but we also coined[14] the expression "dienone–phenol rearrangement", which became the title of my Ph.D. thesis and eventually the accepted generic term in the chemical literature for this type of reaction. My unambiguous demonstration of the course of the methyl migration in the chrysene series (12→13) also seemed to offer independent proof for the correctness of the steroid structure 9a first proposed by Inhoffen. As shown by the structures, such an assumption was unwarranted.

I completed my Ph.D. thesis in 2 years and, shortly before my 22nd birthday, returned to CIBA (which had partially supported my studies at Wisconsin) for another 4 years, during which I resumed work on antihistamines and other medicinal compounds. On the side, I was also able to continue developing my steroid interests, prompted by my Ph.D. thesis, with emphasis on steroid halogenations[15] and dehydrohalogenations[16]—steps that were essential in the early stages of the partial synthesis of the estrogens (see 7→8→6) and proved to be even more important in our subsequent work in the corticosteroid field.

In spite of the industrial setting, I managed to publish a fair amount on my own.[15-17] I considered this activity crucial because I had hoped to establish a sufficient reputation in organic chemistry to permit me entry into academia at some advanced level. This plan turned out to be somewhat naive because at that time moving from industry to academia was a one-way street going in the wrong direction, although a few chemists (e.g., John Sheehan in going from Merck to the Massachusetts Institute of Technology) showed that it was not impossible. At the ripe age of 25, I had accumulated nearly 5 years of industrial experience and concluded that I was ready for an academic career. I had absolutely no luck—very few interviews and no offers—but, in retrospect, this was probably the best thing that could have happened to me because indirectly, it led me much faster into a top university position.

The late 1940s were exciting days in steroid chemistry, especially because the antiarthritic properties of cortisone had just been discovered and I was anxious, therefore, to work on an improved synthesis of cortisone at CIBA. Unfortunately, or really fortunately, permission was not granted because most of that work was conducted at CIBA in Switzerland. Thus, when a chemist friend, Martin Rubin from Schering, who was aware of my restlessness at CIBA, proposed me for an opening as associate director of research at Syntex in Mexico City, I did not reject such a possibility completely out of hand—crazy as it sounded. Not only had I never heard of Syntex, but I knew of nobody doing any chemical research in Mexico except for Russell Marker, who had published a few papers[18] from there. However, when I received an invitation to visit Syntex in Mexico City, with all expenses paid and no other advance commitment, I accepted.

I had never been to Mexico and, as a bonus, decided to include a visit to Havana in my tourist itinerary. George Rosenkranz, the technical director of Syntex (Hungarian-born and Swiss-trained at Eidgenössische Technische Hochschule [ETH] under the famous Leopold Ruzicka), absolutely charmed me personally and professionally. Furthermore, he made me an offer that was really tempting: work in Mexico City on a possible synthesis of cortisone from the steroidal sapogenin diosgenin with half a dozen Mexican collaborators and in laboratories that were amazingly well equipped. Laboratorios Syntex S.A., as it was then called, had a Beckman DU ultraviolet spectrophotometer and the first Perkin–Elmer single-beam IR instrument, which I learned to use in a crash course at Dobriner's laboratory at the Sloan–Kettering Institute in New York. CIBA, in Summit, New Jersey, had no infrared instrument at that time! On the other hand, Syntex had a few rather charming primitive touches. The hoods did not have any exhaust motors (just a pipe leading to the outside), and hydrogenations were conducted in an open courtyard that offered ample space in case of any rupture of the hydrogenation vessel.

The 2 years spent at Syntex working in the laboratory and also supervising the work of a dozen or more Mexican colleagues—notably Octavio Mancera, Jesus Romo, and Jose Iriarte—were among the most productive ones of my chemical career. I had no difficulty in persuading George Rosenkranz that we should publish widely, so as to establish Syntex's scientific credibility, and I became the coauthor of some 60 papers (all of them written by me) on topics that were far from trivial. This was one of the rare instances when the patent applications of a company were governed by its publication policy.

My early interest from Wisconsin and CIBA days in bromination and dehydrobromination led to the initiation of a program on the bromination of Δ^4-3-keto steroids. We showed[19] that under suitable conditions these conjugated ketones (5) could be dibrominated to 2,6-dibromides (14), which upon dehydrobromination provided steroidal 1,4,6-trien-3-ones (15). Aromatization of 15 in the vapor phase in the

I did not meet Russell Marker (second from left) until 1969, when he was honored by the Mexican Chemical Society. On the left is Pedro Lehmann, the son of one of the Mexican founders of Syntex; Roger Adams is standing.

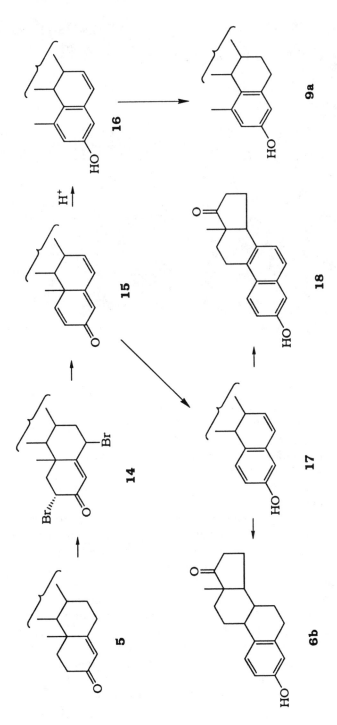

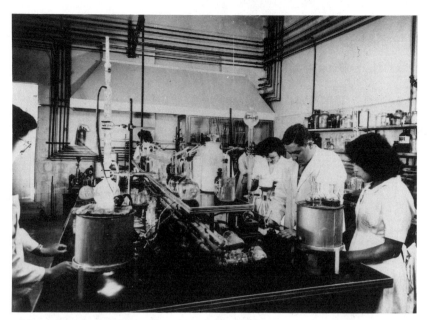

My research laboratory at Syntex in Mexico City in 1950. I am second from the right.

presence of tetralin afforded the 6-dehydrophenol **17** in good yield. When performed in the 17-keto series, hydrogenation of **17** led to estrone (**6b**), whereas selenium dioxide treatment gave equilenin (**18**). In the 17-hydroxy series (**5** = testosterone), this sequence produced estradiol (**6a**) and dihydroequilenin. In other words, we had developed a convenient three- to four-step synthesis of all important estrogens, other than equilin (**23**), directly from the commercially available male sex hormone **5**.

Equilin (**23**) represented an interesting challenge, because it had never been synthesized partially or totally. Only in 1958, on the basis of our extensive work with 19-nor steroids (vide infra), were we able to accomplish a synthesis of this elusive estrogen via 19-nortestosterone (**19**).[20] Bromination at C-6 (**20**) and dehydrobromination yielded the conjugated 4,6-dien-3-one **21**, which was transformed to the nonconjugated isomer **22** and then aromatized by microbiological means.

The availability of the 1,4,6-trien-3-ones (**15**)[19,21] led us also to an examination[22] of the acid-catalyzed dienone–phenol rearrangement of the type I demonstrated[14] in the chrysene series (**12**→**13**). The rearrangement proceeded smoothly and, upon hydrogenation of the C-6—C-7 double bond of **16**, provided an authentic specimen of the 1-methyl-3-hydroxy phenol **9a**, which was different from the specimen

obtained earlier by Inhoffen in the acid-promoted aromatization of the 1,4-dien-3-one (8).

In the same year, Woodward and Singh[23] at Harvard prepared the simple bicyclic dienone 24 and showed that it underwent dienone–phenol rearrangement to the *para*-substituted phenol 26 rather than the *meta* isomer 30, which would be expected on the basis of the santonin (10)→desmotroposantonin (11) precedent. They rationalized these results by invoking a spiran intermediate of type 25b. Because our results in the chrysene (12)[14] and steroid (15)[22] series had demonstrated that an extra double bond in conjugation with the cross-conjugated dienone generates *meta*-substituted phenols (13 and 9a), we postulated that the conjugated carbonium ion 28a leading to the *meta* phenol is favored in this instance at the expense of the higher-energy spiran contributor 28b. Indeed, when we synthesized[24] the bicyclic trienone 27, the predicted[23] *meta* phenols 29 and 30 were isolated. These publications, and many others cited elsewhere,[11] offer a good example of how steroid research has contributed to the solution of mechanistic problems of general organic chemical interest.

Synthesis of Cortisone from Steroidal Sapogenins

The partial aromatization studies described were by no means the end of the various directions in which the original halogenation–dehydrohalogenation work took us. My principal reason for joining Syntex in 1949 was to implement a research program designed to achieve the partial synthesis of cortisone (32) from a readily available plant sterol. Until then, the only route to cortisone (32) was Sarett's 36-step synthesis[12] from the bile acid deoxycholic acid (33). This synthesis was one of the most competitive organic chemical projects of the late 1940s and early 1950s in which many top academic and industrial groups participated. We chose as starting material diosgenin (31), a steroidal sapogenin widely distributed in certain Mexican yams; it served as the commercial starting material for Syntex's production of the male and female sex hormones. In terms of synthetic strategy, the overall conversion of diosgenin (31) into cortisone (32) could be divided into three stages: (1) elaboration of the dihydroxyacetone side chain (32a); (2) generation of the 3-keto-4-ene system (5) of ring A of cortisone; and (3) introduction of the 11-keto function into ring C.

The first problem was the simplest, because the Marker degradation[12] (heating with acetic anhydride, followed by chromium trioxide oxidation and base treatment) converts the spiroketal grouping, encompassing rings E and F of diosgenin (31a), in essentially one step, into the 16-dehydro-20-ketopregnane moiety (34). Various procedures were

DJERASSI *Steroids Made It Possible* 29

available[12] for the further conversion of **34** into the cortisone side chain **32a**, and one of these was described by our group.[25]

Generation of the ring-A substitution pattern was more difficult. In Sarett's synthesis of cortisone (**32**), all intermediates possess the 5β-orientation (A–B *cis* ring fusion). The corresponding saturated 3-ketone **35a** is known[12] to undergo monobromination at C-4 (**35b**); dehydrobromination then leads directly to the required 3-keto-4-ene system (**5**).

35a: X = H
35b: X = Br

5

Unfortunately, intermediates derived from diosgenin (**31**) and many other plant precursors afford 3-ketones (**7a**) of the 5α series (A–B *trans* ring fusion). These intermediates undergo monobromination at C-2 rather than C-4.

Fortunately, our continued interest in steroidal haloketones led to a facile solution[26] that also proved to be of commercial importance. While studying the preparation and behavior of steroidal α-iodoketones, we noted that treatment of the dibromide **7b** with sodium iodide in acetone directly afforded the 2-iodo-3-keto-4-ene system (**36**), which was easily deiodinated with chromous chloride to the desired unsaturated ketone **5**. This method could also be used[27] in the presence of the dihydroxyacetone side chain (**32a**) and proved to be particularly useful in a commercial synthesis of cortisone from hecogenin (vide infra).

The most difficult problem was the introduction of an oxygen function into position 11 of a ring-C-unsubstituted steroid such as diosgenin (**31**). In 1951, several groups—notably at Harvard,[28] Merck,[29]

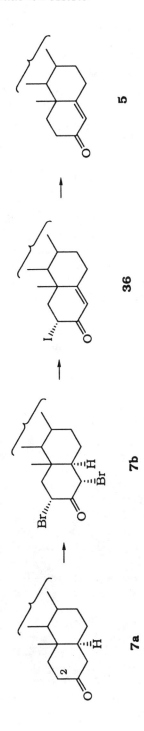

and ours at Syntex—attacked this problem via 7,9(11)-diene intermediates (39), which in turn are readily accessible through the well-known 5,7-dienes (37), followed by catalytic hydrogenation to the 7-ene 38 and mercuric acetate dehydrogenation.[30] Our very first approach,[31] which was performed in both the pregnane and steroidal sapogenin series, was a versatile one that led to both C-11-ketones (42) and 11α-hydroxy-functionalized derivatives. Its key feature was performic acid oxidation to the keto epoxide 40, followed by facile base isomerization to the 11α-hydroxy-8-en-7-one 41.

Once the introduction of the oxygen function at C-11 was solved, judicious combination of all three sequences completed the first synthesis of cortisone from a plant source.[32] The race to synthesize this rare hormone from a source more readily available than bile acids was international.[12] It included two teams (Fieser and Woodward) from Harvard; one (E. R. H. Jones) from Oxford; one (Jeger) from the Federal Institute of Technology in Zurich; and groups from various pharmaceutical companies (Merck being one of the key companies). *Harper's Magazine* in 1951 described the situation in one breathless sentence: "The new ways of producing cortisone come as the climax to an unrestrained, dramatic race involving a dozen of the largest American drug houses, several leading foreign pharmaceutical manufacturers, three govern-

Louis and Mary Fieser in the late 1940s. Photo courtesy of the Harvard University Archives.

ments, and more research personnel than have worked on any medical problem since penicillin."

The race was "won" by our group at Syntex—in the sense of earliest date of submission for publication in 1951[32] in the *Journal of the American Chemical Society (JACS)*. In fact, the same issue of that journal contained the Merck synthesis[33] of cortisone from ergosterol and stigmasterol, and ours[32] from diosgenin. A few months later, we reported a second synthesis of cortisone,[34] this time starting with another steroidal sapogenin, hecogenin (**43a**), which was available from the waste products of Mexican sisal production. In the late 1950s, the British firm Glaxo used this process under license from Syntex for the industrial synthesis of cortisone from East African sisal.

Our synthesis of cortisone from a plant raw material, rather than from animal-derived bile, not only made scientific headlines but even resulted in articles in *Fortune* and *Life*. The *Life* article showed us in a silly picture posed around a bare conference table supporting a huge Mexican yam from which we supposedly made cortisone. George Rosenkranz, the only person in that group over the age of 30, is holding a test tube of cortisone (my vague recollection is that the tube was filled with salt, because we had only made a minute quantity of cortisone at that time), and we seem to be mesmerized by the huge and somewhat ominous-looking yam.

Press conference announcing the first synthesis of cortisone from a plant source at Syntex in Mexico City, 1951. Standing, left to right: A. L. Nussbaum (Department of Biological Chemistry, Harvard), one of my first three Ph.D. students at Wayne; Mercedes Velasco; Gilbert Stork (then at Harvard), consultant to Syntex; J. Berlin; and Octavio Mancera. Seated, left to right: J. Pataki; G. Rosenkranz; Enrique Batres; Carl Djerassi; Rosa Yashin; and Jesus Romo.

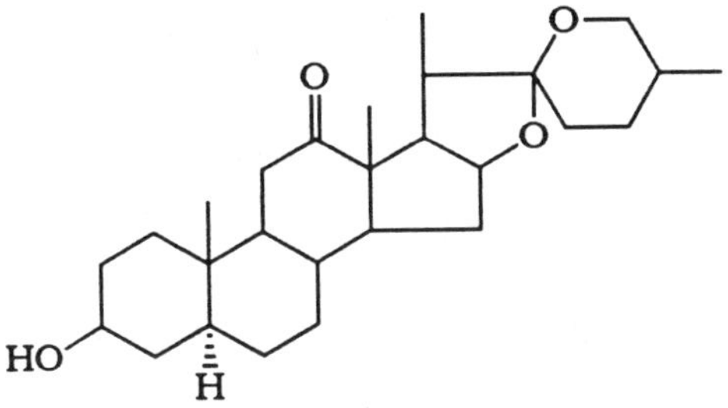

43a

In the summer of 1952, a Gordon Conference focusing exclusively on steroid chemistry was held in New Hampshire at the New Hampton School, a prep school whose regular young inhabitants had strong bladders and took few showers, judging from the extraordinarily high ratio of sleeping quarters to bathrooms. One afternoon, several of the cortisone racers were reminiscing about the events of the past year. Our competition had been forgotten. Proudly and self-contentedly, we recalled that most of the Communications dealing with cortisone had not even been refereed. It may have been R. B. Woodward from Harvard, or maybe Gilbert Stork, who mused, "I bet all it takes these days is the title 'Synthesis of Cortisone' and a reasonable address, and a Communication will appear in the next issue of the *JACS*." That was all the impetus we needed.

Two days later, Stork, Woodward, Sarett (the Merck chemist who had accomplished the *Ur*-synthesis of cortisone from bile acids), and I had concocted a manuscript of less than a thousand words entitled "Synthesis of Cortisone from Neohamptogenin." The authors were F. Nathaniel Greene and Alvina Turnbull; the address, the most prestigious of them all: Converse Memorial Laboratory, Harvard University, Cambridge, Massachusetts. According to that article, our collective nom de plume had made the sensational discovery that New Hampshire maple syrup represented a potentially inexhaustible source of a new steroid sapogenin, now named neohamptogenin, whose chemical structure was that depicted in **43b**.

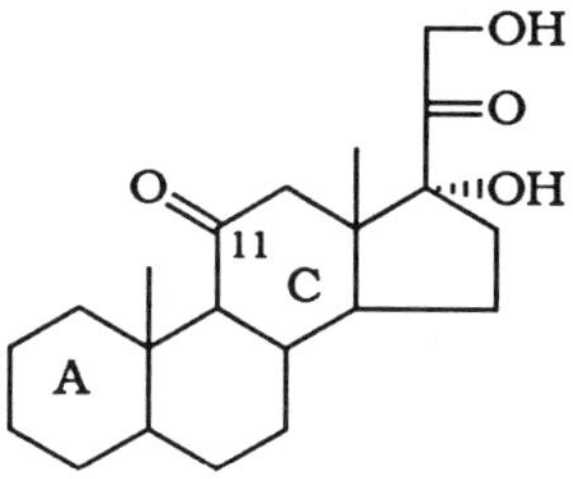

43b

PARTIAL SYNTHESIS OF CORTISONE FROM NEOHAMPTOGENIN

Sir:

Recently,[1] we recorded the isolation of neohamptogenin from the maple

(1) Greene and Turnbull, Trans. Proc. Exp. Ped., 52 (1951) in press.

syrup of a New England species of Agave Klim Linn., and proved its structure as 3-desoxo-4,5-dihydrocortisone (I) (pregnane-11,20-dione-17α,21-diol), the configuration at C-5 not having been determined. We should now like to report the partial synthesis of cortisone from neohamptogenin, which, so far as we are aware, constitutes the first successful introduction of a 3-keto group into an 11-oxygenated steroid devoid of functional groups in ring A.

Neohamptogenin was converted into the cyclic 17,21-sulfite (m.p. 217-220° (dec.), $(\alpha)_D$ +85° (menthol)) in order to protect the dihydroxyacetone side chain, brominated in nitromethane solution to yield 12-bromoneohamptogenin sulfite (not isolated) and then subjected to dehydrohalogenative rearrangement to afford in essentially quantitative yield Δ^7-pregnan-11,20-dione-17α,21-diol sulfite (II) with a characteristic sulfite doublet at 1436 and 1434 cm^{-1}. Dehydrogenation in the usual manner yielded the $\Delta^{5,7}$-diene (III), but attempts to introduce a third double bond with mercuric acetate at 60° for one hour in isopropyl acetate solution containing a trace of iodine- cupric cyanide resulted in the aromatization of ring B and evolution of ca. 70% of the theoretical amount of methane, identified by its Raman spectrum.[2]

(2) N.O. Jawohl, Z. Physik Biochem., 1, 1 (1911).

After a number of unsuccessful attempts, it was found that rearrangement to the $\Delta^{4,6}$-isomer (IV) could be effected in substantially quantitative yield by employing a selective catalyst.[3]

(3) Kindly supplied by Dr. J.S.D.B. Duff of this Laboratory. The details of the preparation of this catalyst will appear elsewhere.

The structure of the $\Delta^{4,6}$-diene (V) was fully established by its ultraviolet absorption spectrum and molecular weight determination (found 417). Oxidative reduction smoothly led to cortisone sulfite from which free cortisone was regenerated on passage through a column (5 x 350 cm.) of Milorganite-S saturated with lead tartrate.

We wish to thank the New Hampshire Maple Growers Association - in particular, Miss Matilda Gill - for a generous supply of neohamptogenin concentrate.

Converse Memorial Laboratory   F. Nathaniel Greene
Harvard University   Alvina Turnbull
Cambridge, Massachusetts

[R. B. Woodward, L. Sarett, G. Stork, C. Djerassi]

The concocted manuscript reporting maple syrup as a source of neohamptogenin.

Gilbert Stork in Paris in 1954. This photograph was taken by W. S. Johnson.

One hardly needs to be a steroid chemist to recognize the unique character of this new, albeit imagined, steroid **43b** from waspish New England rather than the Latino jungles. Neohamptogenin, according to Turnbull, already possessed the crucial oxygen atom in position 11 of ring C—the structural feature of cortisone (**32**) that had presented the greatest obstacle to the Syntex–Harvard–Merck groups, as well as the intact cortisone side chain. What neohamptogenin lacked was an oxygen atom at position 3 of ring A. Therefore, Turnbull and Greene were faced with the task—unprecedented in steroid chemistry in the 1950s (or for that matter even in the 1980s)—of having to introduce an oxygen atom into a "naked" ring A from a precursor that already had the crucial cortisone oxygen atom in ring C. The sum and substance of our communication consisted of a description, all in plausible chemical language with fake literature citations, of how that feat was accomplished.

On the last evening of the conference, Woodward announced that he had been asked by a Harvard colleague to present a preliminary announcement. Woodward's capacity for maintaining a deadpan mien was only surpassed by his ability to metabolize unheard-of quantities of alcohol. He was also a consummate speaker, notorious for his insistence on using chalk (two colors at a minimum) and blackboard, rather than

David Lightner, now a professor at the University of Nevada, then a graduate student, and Keith Brown, now a professor at the University of Campinas, Brazil, then a postdoctoral fellow, under my watchful eye at Stanford in 1961.

slides, thus displaying his virtuosity in drawing chemical structures. Slowly, he led the assembled audience through the step-by-step transformation of our imaginary neohamptogenin into cortisone. Halfway through his lecture, while pontificating on Turnbull's discovery of a new chemical reaction, he was suddenly interrupted by a well-known steroid chemist, a still-running-but-not-yet-arrived cortisone racer, who proclaimed that his group had also performed this reaction. While Woodward threw one of his cold, fishlike, nonblinking stares into the audience, Stork, Sarett, and I were nearly convulsed at this display of competitive priority-manship, quite oblivious to the fact that all of us were infected by that same ubiquitous virus. Only when Woodward approached the very end of his talk and proceeded to explain that Greene and Turnbull had purified the penultimate synthetic intermediate by a passage through a column of Milorganite (a commercial fertilizer), supplied by the New Hampshire Maple Syrup Producers Association, did the audience's shoulders bounce and their faces collapse into laughter.

Upon his return to Harvard, Woodward presented at one of his evening seminars the latest news from the Gordon Conference, including the putative cortisone synthesis from neohamptogenin. I was already back in Mexico City, but Stork reported what transpired. Woodward presented the work in so forceful a manner, the suspense building up so dramatically, that when he concluded, without divulging the identity of the authors, the assembled graduate students and research fellows left with the firm conviction that they had heard exciting new research. Stork was sufficiently irritated at the gullibility of that Harvard audience that after Woodward's departure, he approached a group of students to debunk the chemical tale. Hardly any of the students was prepared to accept that the entire neohamptogenin story was a figment of our collective imagination. It was this event, more than any other, that convinced us not to submit our neohamptogenin communication to the editor of the *JACS*. What if he greased the editorial trail to such an extent that the manuscript might appear in print before we even had time to admit that it was a hoax? Whatever the *JACS* was famous for, humor was not its forte. None of us was willing to be ostracized in the future with our real papers.

This picture, taken at the Prague IUPAC Conference in 1962, looks as if I am testing R. B. Woodward's voice while I. V. Torgov of Moscow listens. Woodward agreed with my interpretation of this photo, which he inscribed "Con amore, Mario."

Although our synthesis of cortisone from diosgenin put Syntex on the international steroid map, none of the chemical syntheses published in August 1951 ever contributed to the treatment of even one arthritic patient. Within months, a group at the Upjohn Company[35] reported an extraordinarily innovative way of achieving the most difficult aspect of the synthesis of cortisone: the placement of the oxygen atom into position 11. They accomplished this feat by introducing microbiological fermentation technology into the steroid field—the first time this had been done industrially—through exposure of the female sex hormone progesterone (**44**) to certain microorganisms, which in one single operation inserted an oxygen atom into position 11 in high yield. The resulting 11α-hydroxyprogesterone (**45**) could be transformed[36] in nine steps to cortisone (**32**), thus making it by far the shortest synthesis of this cortical hormone (14 steps from diosgenin).

44 → **45**

Ironically, the Upjohn discovery of a simplified cortisone synthesis had an enormous commercial impact on Syntex, because in order to exploit its fermentation procedure industrially, Upjohn suddenly needed tonnage quantities of progesterone—amounts that were totally unheard of prior to that time. In 1951, Syntex was the only company in the world that had the technology to synthesize progesterone (**44**) on that scale from yam-derived diosgenin (**31**), and Syntex thus suddenly found itself to be the key supplier of the new raw material for cortisone synthesis.

First Synthesis of an Oral Contraceptive

In 1973, I received the National Medal of Science from President Nixon. (Shortly thereafter, I read in the newspapers that my name was included on Nixon's "White House enemy list" of Watergate fame, probably because of my active participation in the McGovern campaign.) The citation read: "In recognition of his major contributions to the elucidation of the complex chemistry of the steroid hormones and to the application of these compounds to medicinal chemistry and population control by means of oral contraceptives." In actual fact, the work was done 22 years earlier at Syntex in Mexico City—in the same year in which we accomplished the cortisone syntheses from diosgenin (31) and hecogenin (43a).

When I was informed that I would receive the National Medal of Science from President Nixon—hardly a hero of mine—I decided firmly not to be caught smiling at the President when our picture was taken. However, I completely misjudged the rather inane content of our brief conversation as Richard Nixon shook my hand and presented me with the medal.

"How's Stanford going to make out against Cal?" he asked, not knowing that he was speaking to one of the few Stanford professors who had never attended a "big game" between the two football rivals.

National Medal of Science Ceremony at the White House in 1973 on the day that Vice-President Agnew's resignation was announced and two weeks before my name appeared on the White House Enemies list of Watergate fame.

42   PROFILES, PATHWAYS, AND DREAMS

I have to give Nixon credit for trying to make his conversational partner feel comfortable. When he heard me hem and haw, he switched to chemistry. "You know, I never took chemistry at Whittier," he confided. "I got an A in high school, but I never understood it."

I could not help laughing, and that was precisely when the flashbulbs popped. Every one of the big color photographs mailed to me by the White House, including the one inscribed "To Carl Djerassi with best wishes, Richard M. Nixon," had me beaming at the President as if I had been the winning quarterback of the Stanford football team and he the coach. That framed picture is now hanging on the wall of my office, accompanied by an explanatory statement written in beautiful calligraphy by one of my students. It reads "Support Your Local Enemy."

Another and, in many respects, even more enjoyable incident—also associated with oral contraceptives—occurred 1 year later when I

At a Columbia University commencement in 1974 with Frances Hoffman, who was then director of laboratories of Columbia's chemistry department. Her chemical career was launched when she graduated from Mount Holyoke College and became my assistant at CIBA in the late 1940s. Therefore, I take full credit for having made possible the exceptionally well-run operation of that department during Frances's regime.

received an honorary doctorate from Columbia University. In his citation, the president of Columbia alluded to the fact that the most significant impact of our oral contraceptive research had been on the emancipation of women. He had barely finished that sentence when the entire student body of Barnard, the women's college of Columbia, interrupted the president's address by rising in unison and shouting "Yeah!" As soon as the women had calmed down, a second cresting human wave arose: the graduating seniors of the then all-male Columbia College. "Yeah!" they thundered, fisted right arms thrust in the air.

By 1950, the multiple biological functions of the female sex hormone progesterone (**44**) were well known; among them, maintenance during pregnancy of the proper uterine environment and inhibition of further ovulation. Accordingly, progesterone could be considered "nature's contraceptive", and some suggestions had appeared in the literature in the preceding three decades—notably by the Austrian endocrinologist Ludwig Haberlandt—that progesterone might be useful in fertility control.[37] Were it not for the fact that the natural hormone displays only weak activity when given by mouth, it is conceivable that progesterone might have found practical application as an oral contraceptive. Instead, its use in medicine at that time was solely for the treatment of various menstrual disorders and as an occasional palliative for certain types of miscarriage.

Until the late 1940s, the dogma then in vogue,[38] but based on relatively limited data, was that progestational activity was extremely structure specific and restricted only to the natural hormone **44** and a few analogs with a double bond between the C-6–C-7 or C-11–C-12 positions. This supposedly severe structure specificity was supported by the report[38] that even stereoisomers such as 17-isoprogesterone (**46**) or 14-iso-17-isoprogesterone (**47**) were inactive. No wonder that I was impressed as a graduate student when, in 1944, Ehrenstein[39] reported the complicated multistage transformation of the cardiac aglycone strophanthidin (**48**) in 0.07% yield into an oily "19-norprogesterone" (**49**),

46 (14α)
47 (14β)

which, when tested in two rabbits,[40] exhibited the same biological activity as natural progesterone (44). Structurally, 49 differed from the natural hormone only in the replacement of the angular methyl group attached to C-10 by hydrogen, but stereochemically, there were several problems. The configuration of the C-10 hydrogen was unknown and possibly a mixture; the stereochemistry at positions 14 and 17 was predominantly of the 14β,17α type (49)—in other words, compound 49 is a lower homolog of the progestationally inactive 14-iso-17-isoprogesterone (47).

The likelihood that Ehrenstein's structural alteration—the removal of the angular methyl group—was responsible for the high biological activity[40] became more remote when, in 1950, Birch[41] described the synthesis of 19-nortestosterone (19). This substance was identical in every stereochemical detail with the natural male sex hormone testosterone (5) and only lacked its angular methyl group attached to C-10. The key step in Birch's synthesis was the reduction with sodium in liquid ammonia of estrone glyceryl ether (50b) to the dihydroanisole 51. Mild treatment with acid resulted in cleavage of the enol ether without double-bond migration (52); stronger acid or base treatment provided the desired conjugated ketone 19-nortestosterone (19). According to Birch,[42] removal of the angular methyl group resulted in a marked *reduction* in androgenic activity compared with that of the parent hormone.

Nevertheless, in view of our interest at that time[43] in the synthesis of progestationally active steroids, we decided to undertake the synthesis of authentic 19-norprogesterone (55) with the correct stereochemistry at all centers. Fortunately, in 1950 we had synthesized[44] the aromatic analog 53a of progesterone by the earlier discussed bromination–dehydrobromination–partial aromatization sequence (*see* 7→8→6). It proved to be a relatively simple matter to reduce[45] its methyl ether 53b to the dihydroanisole 54 by a modification—lithium in

DJERASSI *Steroids Made It Possible*

Picture taken at the Gordon Conference on Steroids and Related Natural Products in the summer of 1953. The second row (seated) is full of steroid friends: M. Ehrenstein, W. S. Johnson, L. H. Sarett, R. B. Woodward, I. Bush, D. H. R. Barton, Carl Djerassi, K. Heusser, E. R. H. Jones, and O. Jeger.

50a: R=CH$_3$
50b: R=glyceryl

liquid ammonia, developed by my former thesis advisor, A. L. Wilds[46]—of the original procedure (which we had named the "Birch reduction" in recognition of its discoverer). Acid cleavage and oxidation of the alcohol function at C-20 afforded in good overall yield crystalline 19-norprogesterone (55), which possessed the same stereochemistry at all asymmetric centers as natural progesterone (44). Most dramatically, our isomer was shown[47] to be four to eight times as active as the natural hormone—the most potent progestational hormone known at that time. For us, this result opened the floodgates to 19-nor steroid synthesis.

Before synthesizing the 19-nor analogs of other steroid hormones, such as 19-nordeoxycorticosterone (56)[48] and 19-norhydrocortisone (57),[49] we pursued the therapeutically important lead provided by the high progestational activity of 19-norprogesterone (55). For this purpose, we focused on the removal of the angular methyl group in 17α-ethynyltestosterone (ethisterone) (58). Chemically, this substance is a close relative of the male sex hormone testosterone (5), whose androgenic activity (according to Birch[42]) is apparently reduced when its methyl group at C-19 is eliminated (19). However, biologically, 58 should be considered a progestin because of Inhoffen's observation many years earlier[50] that orally administered 17α-ethynyltestosterone (58) was an effective progestational agent, though less active than the parent hormone progesterone (44) administered by the parenteral route. Because the biological potency of progesterone is greatly augmented

Arthur Birch lecturing in the United States in 1954.

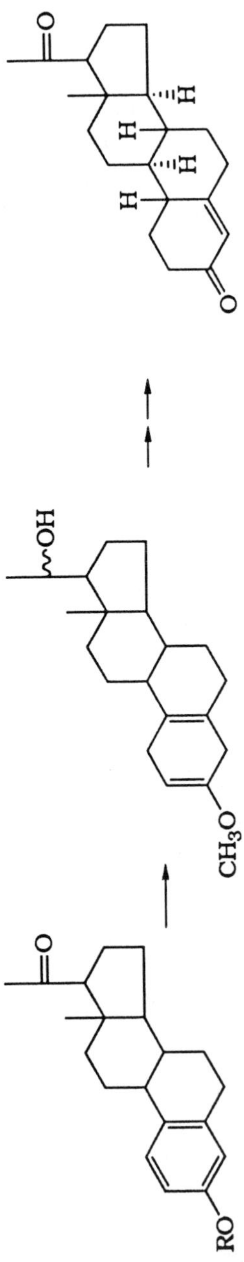

53a: R=H
53b: R=CH₃

54

55

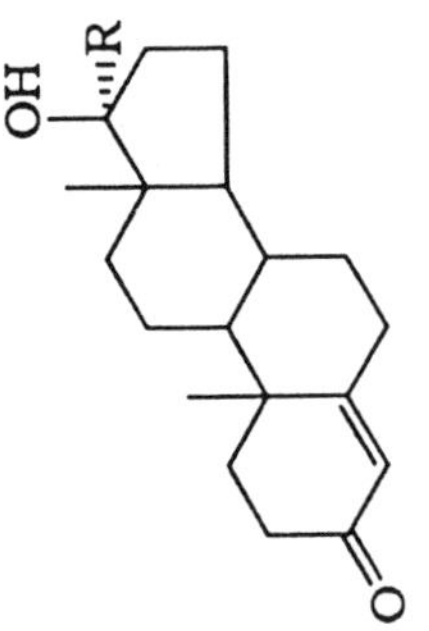

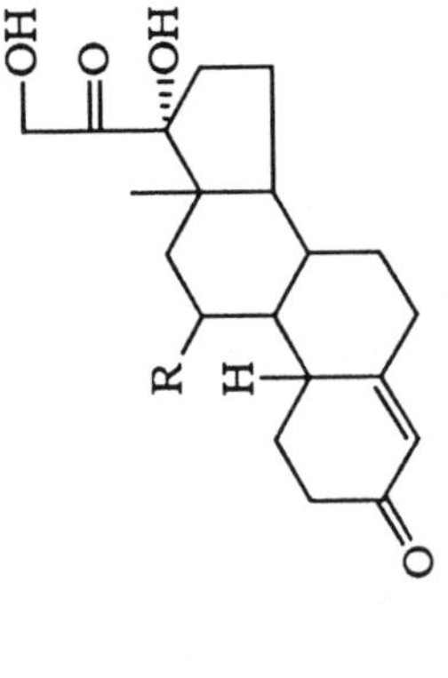

56: R=H
57: R=OH

58: R=C≡CH
5: R=H

DJERASSI *Steroids Made It Possible* 49

upon removal of its angular methyl group (55),[45,47] we argued that a similar change in 17α-ethynyltestosterone (58) should result in increased oral activity.

We accomplished the first synthesis of 19-nor-17α-ethynyltestosterone (62)—now generically known as norethindrone—by oxidation of 19-nortestosterone (19) (obtained by the Wilds procedure[46] from estrone methyl ether 50a) to the corresponding C-17 ketone 59, protection of ring A via the enol ether 60, ethynylation at C-17, and acid cleavage of the resulting intermediate 61. This first synthesis of what eventually proved to be a synthetic oral contraceptive was completed on October 15, 1951, by a young Mexican chemist, Luis Miramontes, who was working on the equivalent of a B.S. thesis in our laboratory.

We immediately submitted the substance for biological evaluation and found it to be the most potent orally effective progestin described up to that time. Our Mexican patent application was filed on November 22, 1951[51]; the invention is now included in the National Inventors Hall of Fame in the U.S. Patent Office. I reported the chemical results, together with the substance's biological activity, at the April 1952 national meeting of the American Chemical Society in Milwaukee.[52] The full article, with complete experimental details, appeared 3 years later.[53] We supplied norethindrone (62) to various investigators: to Hertz[54] (National Cancer Institute) and Pincus[55] (Worcester Foundation) for further biological scrutiny and to a group of clinicians[56] for human trials. In fact, the late Edward Tyler of the Los Angeles Planned Parenthood Center merits attention for having reported in November 1954[56] the first clinical results with norethindrone for the treatment of various menstrual disorders and fertility problems, long before any other clinical studies with 19-nor steroids.

On August 31, 1953, well over a year after our first public report,[52] Frank B. Colton of G. D. Searle & Company filed a patent[57] for the synthesis of the β,γ-unsaturated analog 63 (now known generically as norethynodrel) of norethindrone (62). As expected (see 51→52→19), mild treatment with acid, or indeed just human gastric juice,[58] converts Searle's norethynodrel (63) into Syntex's norethindrone (62). Syntex's

Induction into the National Inventors Hall of Fame for synthesis of the first oral contraceptive. Secretary of Commerce Juanita Kreps is sitting in the background.

patent was thus circumvented in the bottle or pill, though not in the stomach; for reasons described elsewhere,[59] the question of whether synthesis in the stomach is an infringement of an issued patent was never tested in court.

Both norethindrone (62) and norethynodrel (63) were examined in animals by Pincus et al.[55] for ovulation inhibition and found to be highly active by the oral route. Pincus, who was a consultant for Searle, subsequently concentrated on Searle's norethynodrel, while Syntex made arrangements with other companies, because at that time the Mexican company had neither pharmaceutical outlets nor biological laboratories. The first distributor that Syntex chose was Parke–Davis, and the first formal approval by the Food and Drug Administration (FDA) for the clinical use of norethindrone (62) under the trade name Norlutin for the treatment of menstrual disorders was issued in 1957— the same year that the FDA approved Searle's norethynodrel (63) for similar clinical indications.

At this point, the two paths diverge in terms of how these two substances were introduced into medical practice as oral contraceptives. This story has already been told elsewhere[59] and will not be repeated here. It is interesting to note that 35 years after its first synthesis, and a quarter of a century after the FDA's approval of its use as an oral contraceptive, norethindrone (62) still represents one of the two most widely used steroid contraceptives in the world. The other one is Wyeth's norgestrel (64),[60] which differs from 62 in the nature of its C-13 substituent (ethyl rather than methyl). Norethynodrel (63) is essentially not used any more.

64

Accomplishing all this steroid research in just a couple of years would have been noticeable anywhere; doing it in Mexico City at that time was so unexpected that this work received international attention. It also produced the first (and only) offer of an academic job—a tenured associate professorship from Wayne University (now Wayne State University) in Detroit. Although hardly a top university when compared with the Ivy League schools, the University of California campuses, or California Institute of Technology, Wayne University

With Alejandro Zaffaroni, who was then (1959) executive vice-president of Syntex in Mexico City.

nevertheless provided a crucial stepping stone, just as it had done a few years earlier for H. C. Brown, who initially had also found it difficult to secure an academic position.

In January 1952, I left sunny Mexico City and drove to cold, dirty, slushy Detroit. In the end, it was the most direct route to California, where I have now spent three decades.

Optical Rotatory Dispersion and Circular Dichroism

The 5 years at Wayne University were important; they represented my first academic job, and a lot of new work was started there. My research laboratories were awful—my group was housed in the oldest building of the university, and my students had to cross one of the busiest streets (four lanes, one-way traffic) to get to the stockroom. One of my students may have been the only chemist in America to have received a ticket for jaywalking while running to the stockroom on a rainy day.

However, the stockroom was first class and generously stocked in terms of chemicals and glassware; we had good IR and UV equipment; and, most importantly, the university and departmental administrations were highly supportive. Aside from continuing various steroid

Huang Liang, my first Chinese postdoctoral fellow, was the chemist at Wayne who received the jaywalking ticket while crossing the street from our laboratory to the stockroom. When I met her again in 1973 in Beijing, where we are shown eating lunch, we reminisced about this unique characteristic of doing chemical research in Detroit.

projects, my students started on a major line of natural-products chemistry in the area of antibiotics,[2] alkaloids,[4] and terpenoids.[3] In addition, shortly after my arrival at Wayne, I was able to get one of the first spectropolarimeters constructed by the O. C. Rudolph Company and to initiate research on chiroptical methods. For instrumental reasons, we commenced with optical rotary dispersion (ORD), the laborious manual point-by-point wavelength measurements being taken by the wives of my graduate students, notably Rosemarie Riniker, Tita Halpern, Avrill James, and Betty Mitscher. In typical male fashion, all the photographs in my possession show pictures of the husbands, but not of the wives!

Determination of Absolute Configuration by Optical Rotatory Dispersion

Those were the days when D. H. R. Barton[61] and, shortly thereafter, W. Klyne[62] in England demonstrated the utility of molecular rotation differences (at a single wavelength corresponding to the sodium D line) in the steroid field. My own experience[17] with that method had given me the intuitive feeling that measurements of optical rotation at different wavelengths (i.e., optical rotatory dispersion curves), especially in the ultraviolet, would be more informative than just focusing on a single wavelength, but I had been unable to test this assumption for lack of suitable instrumentation.

Only when the Rudolph spectropolarimeter arrived at Wayne, could we undertake some experimental explorations. Luckily, I made the best possible choice for experimental substrates by selecting steroid ketones. The low absorption of the carbonyl group in an experimentally accessible region of the ultraviolet spectrum permitted ORD measurements through and beyond the $n \rightarrow \pi^*$ absorption maximum around 280 nm. The steroid framework turned out to be the ideal structural setting in which to test my belief that the shape or sign of the ORD curve would be diagnostic for the location of the carbonyl group. Our first few papers[63] confirmed this hypothesis. More extensive measurements among steroids and related polycyclic natural products then demonstrated that the predominantly bicyclic environment around the carbonyl group reflected itself in the sign, amplitude, and shape of the ORD curve.

With my first Ph.D. student, Carl Lenk, at Wayne University, Detroit, in 1952.

This conclusion was established[64] through extensive syntheses of optically active bicyclic ketones of known absolute configuration and comparison of their ORD curves with those of their tetracyclic steroid counterparts. For instance, the *trans*-9-methyl-3-decalone with the absolute configuration denoted by structure 65 displayed a negative Cotton effect in its ORD curve that was the mirror image of the positive Cotton effect associated[63] with 3-ketosteroids of the 5α configuration (e.g., 66). The bicyclic ketone 65 was actually synthesized from a precursor of known absolute stereochemistry. However, if the absolute configuration of 65 had not been known, just comparing its ORD curve with that of the appropriate steroid reference ketone (i.e., 66) would have demonstrated that its absolute configuration was opposite to that of the steroid ring A—B decalone system.

65

66

Through numerous ORD comparisons of steroidal and terpenoid ketones with a variety of specially synthesized chiral ketones of known absolute configuration, we laid the foundation for the use of optical rotatory dispersion (and later optical circular dichroism) in the determination of absolute configurations[64] or detection of conformational changes[65] in organic molecules. Our concurrent structural work in the terpenoid field immediately provided striking examples of the power and simplicity of this chiroptical approach to the determination of absolute configurations.

One of the great triumphs of "paper biogenesis," which subsequently prompted an enormous amount of experimental biosynthetic work, was the isoprene rule in the steroid and terpenoid fields. In his famous 1953 paper,[66] Ruzicka pointed out that of the many sesquiterpenes known to follow the isoprene rule, none had been encountered with the bicyclic skeleton 67. He concluded that "this appears to indicate that the biogenesis of the steroids, diterpenes and triterpenes differs in some fundamental detail from that of the monoterpenes and sesquiterpenes."

Within a year of that conclusion we demonstrated[67,68] that the sesquiterpene iresin, which we isolated from a Mexican plant, possessed structure 68 (without stereochemical implications) and hence

Leopold Ruzicka in Zurich on February 14, 1951. This photograph was taken by W. S. Johnson.

67

68

represented the long-sought structural "missing link" between the higher and lower terpenes. Because iresin links the higher and lower terpenes, the question of its absolute configuration was highly relevant. To our great surprise, we found[69] that several iresin degradation prod-

A Djerassi group at Wayne University in 1955. Left to right: H. Bendas (Israeli postdoctoral student, now deceased); Cecil Robinson (Johns Hopkins); Albert Bowers (CEO, Syntex); J. A. Henry (postdoctoral student from Scotland); J. Grossman; Gerald Krakower (Bar Ilan University, Israel); F. Donovan (postdoctoral student from Australia); John Zderic (vice-president, Syntex); Jack Fishman (Rockefeller University); R. Hodges (Department of Agriculture, New Zealand); Carl Djerassi; H. S. Monsimer; S. C. Pakrashi (director, Indian Institute for Experimental Medicine); G. R. Pettit (Arizona State University); and B. Riniker (Ciba–Geigy).

ucts, such as the ketones **69** and **70**, had Cotton effect curves of the opposite sign as those of the two steroid reference ketones **71** and **72**. Therefore, iresin has the "wrong" absolute configuration (as denoted in **68**) compared with the reference steroids. Since our elucidation of the stereostructure of iresin (**68**), numerous sesquiterpenes have been isolated that also contain the isoprene-derived skeleton **67**. Several of these also possess a "nonsteroid" absolute configuration.

69   **70**

               71   72

The Octant Rule

We then extended our optical rotatory dispersion measurements to my original love from graduate school days—the steroidal α-haloketones.[70] With William Klyne, we made the crucial discovery[71] that an axial α-halogen substituent controlled the sign of the Cotton effect of the ketone. We called this observation the α-haloketone effect and pointed out that this afforded a means of establishing the absolute configuration of cyclohexanones without requiring a reference compound (such as a steroid) of known absolute configuration. Shortly thereafter, this α-haloketone rule was shown to be a special case of the octant rule.[72] The article introducing this rule has turned out to be one of the most-cited papers[73] in my bibliography.

I still recall the day in 1958 when I presented a seminar at Harvard and had a long session in R. B. Woodward's office with William Moffitt and his graduate student Albert Moscowitz (now a professor of physical chemistry at the University of Minnesota). Moffitt and Moscowitz had been working on theoretical aspects of optical activity, and Woodward had called to their attention our experimental ORD work and especially our α-haloketone rule. In an exciting session lasting several hours, acted out by Moffitt and Woodward in the photograph on p 60 that I took in Woodward's office, we realized that the octant rule represented an important extension of the α-haloketone rule. The connection was clear because every prediction by Moffitt and Moscowitz was immediately supported by my producing some experimental verification. The octant rule explained virtually all of our published and unpublished results. It provided a superb and very rapid method for establishing absolute configurations of ketones without resort to standards of known absolute configuration. There is little doubt that work in this field would have been delayed by many years if steroid ketones had not been used fortuitously by my group as the initial test substrates.

In our original version of the octant rule,[72] the cyclohexanone model was divided into three planes (Figure 1) corresponding to the

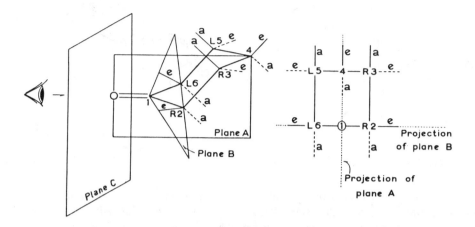

Figure 1. The octant rule for predicting the sign of the optical rotatory dispersion Cotton effect of a substituted cyclohexanone. (Reproduced with permission from ref. 81. Copyright 1960 McGraw–Hill.)

W. Moffitt (left) and R. B. Woodward clowning in Woodward's office in 1958, with the octant rule diagram in the upper right corner of the blackboard.

A Djerassi group at Wayne State University in 1957. Seated, left to right: Jim Kutney (University of British Columbia); Robert Engle; Ami Kapoor; Ben Gilbert (Codetec, Brazil); Howard Smith (Vanderbilt University); Peter Kan; and Gerald Krakower (Bar Ilan University). Standing, left to right: Rolf Mauli (Ciba–Geigy); Kay Thieberg; Jeanne Osiecki; Larry Geller (New England Nuclear); Jim Gray; Pete Eisenbraun (Oklahoma State University); Carl Djerassi; Sumner Burstein (University of Massachusetts); Riccardo Villotti (Pierrel, Milano); Lester Mitscher (University of Kansas); Otto Halpern (former director of chemical development, Syntex); and Peter Lemming (Chas. Pfizer, Ltd., UK).

nodal and symmetry planes of the 280-nm transition of the carbonyl chromophore. These three planes create eight octants, and the presence of substituents in each octant is given a qualitative rotational contribution. Substituents in the planes (i.e., substituents attached to C-4 as well as the equatorial groups attached to R2 and L2 in Figure 1) are assumed to make negligible contributions, while the axial substituents adjacent to the carbonyl group have the greatest effect and behave exactly as predicted by the α-haloketone rule:[71] axial substituents attached to L2 make a negative rotatory contribution, whereas an axial

Carl Djerassi (without safety glasses!) with a laboratory assistant in Mexico City in 1951, during the final stages of the Syntex cortisone synthesis.

R2 group contributes in a positive sense. A further consequence of the octant rule is that it takes into consideration substituents at the two β-carbon atoms (L3 and R3). The axial and equatorial substituents at carbon atoms R3 are assumed to make a negative rotatory contribution; a positive one is expected from axial or equatorial groups attached to L3. Refinements during the next two decades—notably by one of my former students, D. A. Lightner (now a professor at the University of Nevada)[74]—have shown that front-octant effects, as well as those associated with axial substituents in the β position, have to be evaluated in a more sophisticated manner. Nevertheless, there is little doubt that the original version of the octant rule has led to hundreds of applications in organic stereochemistry and conformational analysis. In many cases, if the absolute configuration of the ketone is known, its conformation can be determined by the octant rule. Conversely, if the conformation of the substance is established, it can then be assigned the correct absolute configuration by a consideration of the sign of its Cotton effect.

As an example, the demonstration[75] of conformational mobility in *trans*-2-chloro-5-methylcyclohexanone (73) can be cited. The rule predicts a positive Cotton effect for conformation **73a** (because the corresponding chlorine-free 3-methylcyclohexanone shows a positive Cotton effect) and a negative one for conformation **73b**, whose sign is governed by the axial chlorine atom. In actual fact, the substance displays a negative Cotton effect in octane solution, which was altered to a positive one in methanol. This behavior is in accord with the generalization[76] that an increased proportion of the equatorial α-haloketone **73a** can be expected in the polar medium (methanol) at the expense of the axial conformer **73b**, which is favored in the nonpolar medium (octane).

73a ⇌ **73b**

In 1957 I took a 2-year leave of absence from Wayne State University to return to Mexico City as research vice president and board member of Syntex—a company for which I had served as a consultant in the intervening 5 years. Syntex had just been sold by its Mexican owners to an American investment banking firm and was about to go public, a move that would permit a substantial growth in its research activities. It was difficult to resist the temptation of implementing this expansion. Even more important was the great physical pain in my knee from which I suffered during Detroit winters. In 1957, I was living on 24 aspirins per day and had to use crutches. I decided to go to Mexico City for a major operation: a permanent knee fusion to be performed by an internationally known Mexican surgeon, who himself had undergone such an operation. Somehow, I felt more comfortable with a surgeon who knew from personal experience what it meant to live with a fused knee joint. The operation was so successful that I have since resumed skiing, even though this involved developing a new "fused-knee" technique.

I brought with me to Mexico City some of my graduate students and postdoctoral fellows from Wayne, among them A. Bowers (an Englishman, now chief executive officer of Syntex), J. Kutney (a Canadian, now a professor at the University of British Columbia), R. Mauli (a Swiss, now at Ciba–Geigy), R. Villotti (an Italian, who thereafter became Italian manager of Syntex), J. S. Mills (an Englishman, now chief chemist of the National Gallery, London), O. Halpern (a Costa Rican and eventually director of process development for Syntex), P. Crabbé (a Belgian, who later directed the Mexican research program of Syntex), and J.

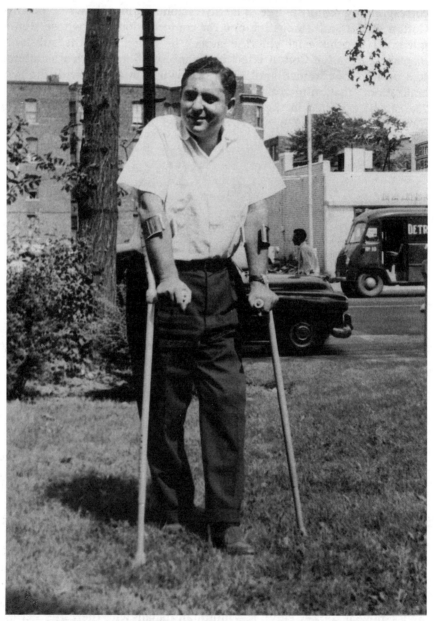

At Wayne State University in 1957, when I decided on a permanent knee fusion after having been sentenced to living with crutches and a brace as a result of my earlier skiing accident.

Zderic (an American and currently a vice president at Syntex). This group, together with H. J. Ringold from the University of Washington and a few Mexican Ph.D.-level chemists, notably J. Iriarte, O. Mancera, and J. Romo, provided a marvelous international flavor. We performed a series of investigations in the androgen–anabolic,[77] corticoid,[78] and progestational[79] areas that, according to a 1957 Gordon Conference talk by L. F. Fieser, made Syntex in the 1950s the preeminent steroid laboratory in the world in terms of scientific publications. In addition, we continued our ORD work with special emphasis on conformational problems associated with halogenated steroid ketones.[80] Among the commercially important products (e.g., norethindrone) developed at that time by Syntex, the topical corticosteroids synthesized with my ex-Wayne group[78] eventually became real money winners under the trade names Synalar and Neosynalar.

My 2-year leave of absence from Wayne was supposed to end in the fall of 1959, at which time I had also completed my first book, which summarized all of our ORD work.[81] An amusing aspect of my publishing contract with McGraw–Hill is worth recalling. When the publisher invited me to prepare the first book dealing with organic chemical applications of optical rotatory dispersion, I insisted on a special condition. I felt that such a book, consisting primarily of very recent advances from our own laboratory in a rapidly developing field, ought to be published with minimum delay. Therefore, I asked for a penalty clause, whereby my royalties would escalate by 1% for each week that the book's appearance would be delayed beyond my 6-month deadline, but as a quid pro quo I offered a similar royalty reduction for each week's advance appearance prior to my requested publication date. To everyone's surprise, the McGraw–Hill lawyers accepted my proposal, provided I agreed to return the corrected page proofs from Mexico City within 24 hours of their receipt. This condition was supposed to prevent a horror scenario, whereby my royalties might escalate to unprecedented heights by my simply sitting on the page proofs! As a final compromise, the publisher set each chapter in print as it was received, rather than waiting for the entire manuscript. I managed to finish the book in approximately 3 months by sticking to a very rigid Monday–Wednesday–Friday writing schedule. In the end, the book appeared exactly on time without any royalty concessions by either party.

My knee fusion immobilized me in Mexico City for almost 6 months, but after that time I traveled to Detroit every 8 weeks, in addition to maintaining intimate contact by long-distance telephone with all of my graduate students at Wayne State University. Although that interval of telephonic research direction was caused by my physical immobilization during the long recovery phase, over the years fairy tales have been told that this was my usual modus operandi. By now these stories have reached the hallowed stage of *se non è vero, è ben trovato*.

In early 1959, my former University of Wisconsin professor, W. S. Johnson, asked if I would be interested in joining him if he were to

I am often asked whether I would consider another operation to convert my fused left knee into a bendable facsimile, given present advances in artificial joints. The answer is a categoric no. Except for jogging and tennis, I have learned over the course of 30 years to cope well with a stiff knee—so much so that I consider it peculiar that other people can bend both their legs. Not only have I developed a skiing technique for stiff-legged persons, but I also enjoy the fringe benefits that go with a fused knee: first-class plane travel and excellent opera and theatre seats. As seen in this picture from a 1990 ski excursion in the Sierras, I have even been known to show off in public.

move to Stanford University as the new head of its Chemistry Department. A joint visit by the two of us to Palo Alto, followed by a lengthy conversation with the legendary Stanford provost F. E. Terman, who promised the construction of a new laboratory building for the two of us, convinced both Johnson and me to accept professorial offers from Stanford University. We officially joined the faculty in 1959 but did not actually arrive in Palo Alto until the Stauffer Chemical Laboratory had been constructed in 1960. However, most of my Wayne University research group, headed by my senior research associate E. J. Eisenbraun (now a professor at Oklahoma State University), moved there in the fall of 1959. Eisenbraun represented Johnson and me in the negotiations with the building architects and then contributed greatly to getting the Johnson and Djerassi research groups settled in our new environment.

Early Circular Dichroism Studies

Scientifically, the 30 years spent so far at Stanford University have been important because they covered five new areas of research, in addition to the lines started at Wayne State University. The first was an extension of our earlier ORD work to optical circular dichroism (CD). A key role was played by Edward Bunnenberg, a postdoctorate fellow interested in instrumental improvements of chiroptical methods, who moved with my group to Stanford and became my longest-serving collaborator in this field until his premature death from cancer in 1984. We secured one of the first CD instruments—a prototype constructed by JASCO in Osaka, Japan—which in 1961 enabled us to extend our ORD work to CD measurements of steroids and many other chiral molecules. We did not repeat (in terms of CD) our earlier ORD studies of saturated and conjugated steroid ketones, because such an investigation had been published in 1961 by Legrand and collaborators[82] in France, where some of the early instrumental CD developments had been pioneered.[83] As expected on theoretical grounds, all of our ORD generalizations of steroid ketones were found by the French investigators to be equally applicable to the CD technique. Therefore, we focused on new areas where ORD and CD might provide spectroscopic or conformational insight and on approaches that would extend these techniques to other chromophores.

One of the first examples[84] of our new CD efforts again came from the steroid ketone field. In contrast to the wealth of information on the close correlation of ultraviolet absorption and ORD among α-bromo- and α-chlorocyclohexanones, very little was known about such correlations among α-iodoketones and particularly about the location of the $n \rightarrow \pi^*$ absorption band in such iodoketones. The puzzling aspect

Press conference in my Stanford office in 1961 on the occasion of the opening of the Stauffer Organic Chemistry Building. On the left, bottom to top: Carl Djerassi, E. R. H. Jones (Oxford), and W. S. Johnson. The others are reporters.

was that, whereas halogen-free cyclohexanones display their absorption maximum near 280 nm and undergo a bathochromic shift upon α-bromination or chlorination, introduction of an α-iodo substituent results in a substantial hypsochromic shift to 258 nm in the ultraviolet spectrum and a bathochromic shift in the ORD or CD spectrum. We were thus able to demonstrate that the 258-nm absorption band is due to the iodine atom and is not optically active, whereas the ORD and CD Cotton effects are associated with a second ultraviolet absorption maximum in the 290-nm region. The latter is of sufficiently low intensity as to be masked by the more-intense 258-nm iodine absorption. Hence, it is not observed under the usual ultraviolet absorption measurement conditions. The 290-nm $n \rightarrow \pi^*$ carbonyl transition, however, is optically active and readily detectable by ORD and CD.

A second area dealt with an important extension of the original[72] octant rule (limited to asymmetrically perturbed but intrinsically symmetrical chromophores) through the concept of inherently dissymmetric chromophores,[86] to which Albert Moscowitz had made major theoretical contributions. The work was really an outcome of an earlier investigation with Kurt Mislow (then at New York University and shortly thereafter at Princeton University) on the ORD properties of hindered biaryls.[85] The triangular Stanford–Princeton–Minnesota collaboration was productive, joyful, and often boisterous. (Just imagine three verbal and literate prima donnas, shown in the photograph on p 71, arguing about the wording of each sentence of a joint paper.) We examined and interpreted the CD behavior of a wide variety of β,γ-unsaturated ketones such as dehydronorcamphor (74) or dimethyldibenzosuberone (75),[87] as well as bridged and unbridged biphenyls and binaphthyls.[88] Aside from offering a simple means of assigning absolute configurations to such molecules, this extension of the octant rule allowed the significant dissection of spectral contributions from overlapping absorption bands.

74

75

The Concept of Chromophoric Derivatives

As early as 1959, we asked how ORD might be used for the solution of stereochemical problems among "transparent" functionalities, such as hydroxy or amino substituents. We found that the C=S chromophore was especially suitable in that respect and demonstrated—first by ORD[89] and subsequently by CD[90] techniques—that conversion to the xanthate provided derivatives with strong Cotton effects in an accessible spectral region (>300 nm). Frequently, the sign of the derivative's Cotton effect could be used to assign the absolute configuration of the parent α-hydroxy or α-amino acids, which were the first test substrates.

This concept of chiral chromophoric derivatives was extended to a variety of derivatives, such as acylthioureas,[91] nitroso derivatives,[92] and nitrites.[93] Examples of their utility are the two epimeric 20-hydroxy steroids 76a and 77a, which show only "plain" ORD curves and no CD Cotton effects in the accessible spectral range above 200 nm. The corresponding xanthates (76b vs. 77b) or nitrites (76c vs. 77c), however, display[93] mirror-image Cotton effect curves and thus lend themselves to convenient assignment of stereochemistry. A detailed review of this

76a: R = H
76b: R = CH₃–S–C(=S)-
76c: R = NO

77a: R = H
77b: R = CH₃–S–C(=S)-
77c: R = NO

work on ORD and CD studies of otherwise transparent functionalities was presented in my Centenary Lecture of the Royal Chemical Society.[94]

Variable-Temperature Circular Dichroism for Conformational Studies

As demonstrated in so many of our early studies with cyclic ketones, ORD and CD represent extraordinarily sensitive probes for detecting conformational changes (cf. 73a vs. 73b). This becomes even more significant when dealing with acyclic chromophores (for instance, xanthates 76b and 77b), in which free rotation around the C=S chromophore is also feasible. In an attempt to widen the scope of conformational applications of CD, Keith Wellman (now a professor of chemistry at the University of Miami) and Edward Bunnenberg constructed a cell in which CD measurements could be performed down to liquid-nitrogen temperatures.[95] This technique of temperature-dependent circular dichroism was initially applied to steroids. It permits the calculation of conformer populations and thermodynamic data[96] and has been used extensively in our laboratory and subsequently by many others. Because such low-temperature studies have to be performed in solvent mixtures, we examined with Albert Moscowitz[97]—again using steroid ketones as initial substrates—the effect of asymmetric solvation around a chromophore. With that theoretical information as background, it was then possible to attack a variety of subtle conformational problems by using this variable-temperature CD technique.

For instance, in the middle 1960s the preferred conformation of the C-17 acetyl group in 20-ketopregnanes was unknown, in spite of the relevance of this information for biomechanistic conclusions. A detailed analysis[98] of variable-temperature CD curves of a group of 20-ketopregnanes led to the conclusion that the low-energy conformers can

While discussing a joint publication on circular dichroism at Stanford in the early 1960s. Left to right: Carl Djerassi, K. Mislow, and A. Moscowitz. Photo courtesy of Stanford University News and Publication Service.

be depicted as **78a** and **78b**, with the latter being more stable by approximately 1.1 kcal. Further verification was eventually provided by X-ray crystallographic studies.[99]

In determining the rotamer composition among chromophoric derivatives such as xanthates, we again chose steroids as initial test cases.[100] Thus, 5α-pregnan-20β-ol methyl xanthate (**79**) showed little variation in rotational strength over the range $+25$ to -192 °C, whereas

78

78a

78b

dramatic changes were encountered with 5α-cholestan-3β-ol methyl xanthate (**80**). Because it is reasonable to assume that the Me–S–C(=S)–O– bonds already exist in the energetically preferred orientation, the chief variable ought to be possible rotation about the bond connecting the xanthate oxygen to the steroid.

In the 20-hydroxy xanthate **79**, free rotation around the C-17–C-20 single bond is severely impaired because of steric interaction of the C-21 methyl and xanthate groupings with the C-18 angular methyl substituent. Therefore, one would expect the most stable rotamer to already predominate at +25 °C. The most likely candidate, in which all skew interactions are minimized, is represented by the Newman projection **79a**, in which the system is viewed from C-17 toward C-20.

On the other hand, in the 3-hydroxy xanthate **80**, only slight, if any, impediment to free rotation should exist. Therefore, at +25 °C the extrachromophoric influences—while reflecting contributions from all possible rotamers—might be very different from those experienced by the chromophore at −192 °C, when it is in its most stable rotameric form. Such an interpretation was confirmed by the 1400% change in rotational strength of **80** in going from +25 to −192 °C.

Isotopically Engendered Chirality of Ketones

A very different application of CD, which again had found its first impetus in our laboratory among ketones, is represented by the chiroptical behavior of molecules that owe their chirality to isotopic substitution. Optical rotation measurements at the sodium D line have been measured for such compounds since 1949 (e.g., 81[101]), but their rotations are extremely small.[102] We felt that a significant test case would be chiral 3-deuteriocyclopentanone (82), which is readily accessible[103] in six steps from (+)-α-pinene, because asymmetric disturbance of the carbonyl chromophore in the cyclopentanone series is usually reflected in particularly large Cotton effects.[104] Whereas early ORD measurements[103] failed to demonstrate any rotatory contribution, repetition of this work by CD[105] uncovered the existence of a negative Cotton effect, even though a positive Cotton effect had been predicted for the (R)-antipode 82, on the basis of the tenets of the original octant rule.[72] This apparent anomaly prompted a much broader investigation of the synthesis and chiroptical properties of ketones with isotopically engendered chirality. The topic was synthetically challenging, raised interesting theoretical questions, and at the same time provided insight into the conformational preference in conformationally mobile, isotopically substituted ketones: for example, is axial deuterium preferred over axial hydrogen?

We proceeded systematically by synthesizing from terpenoid precursors of known absolute configuration the four chiral monodeuterated cyclohexanones 83–86[106,107] with bulky substituents in the 4-position, which would lock the cyclohexanone into one chair conformation. Circular dichroism measurements of these substances, as well as of several others synthesized in our and other laboratories, defined the rotatory contribution of axial and equatorial deuterium in the α and β positions of the cyclohexanone ring: in general, deuterium produces a dissignate (i.e., antioctant) effect. Similar syntheses of "locked" chiral gem-dimethyl-substituted cyclohexanones (e.g., 87 vs. 88)[108] or cis-3,5-dimethylcyclohexanone (89),[109] in which one of the methyl groups was

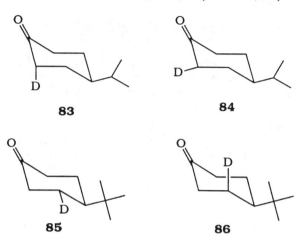

replaced by a trideuteriomethyl analog, provided data for the sign (again dissignate) and magnitude of the rotatory contribution of axial and equatorial trideuteriomethyl substituents compared with their hydrogen analogs.

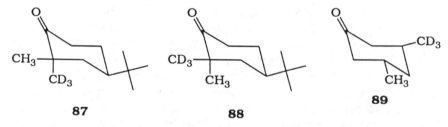

With this basic information, we could now examine some of the conformational questions by variable-temperature CD measurements. The simplest example is optically active 3-deuteriocyclohexanone,[110] with its dynamic conformational equilibrium between the two chair conformations with deuterium located in the equatorial (90a) and axial (91a) positions. However, careful CD measurements[106,107,110] did not reveal any differences in the amplitudes of the Cotton effect from +25 to −190 °C. This negative result is not too surprising, because the energy difference between 90a and 91a is probably on the order of a few calories per mole, and the temperature-induced equilibrium shift is equally small—perhaps less than 1% over the accessible temperature range. Furthermore, because the Cotton effect amplitudes of the two conformers are themselves small numbers, any variation of the CD amplitude with temperature may well be below the limit of experimental detectability. Therefore, only in those situations where the rotational strengths of the participating conformers are large numbers of opposite sign can one hope to detect the presence of small energy differences of the type introduced by isotopic substitution.

DJERASSI *Steroids Made It Possible*

90a: R=H
90b: R=CH$_3$

91a: R=H
91b: R=CH$_3$

Use of Chiral Probes for Conformational Studies

We solved this problem by the introduction, adjacent to the carbonyl chromophore, of a *gem*-dimethyl substituent.[111] The methyl groups, per se, do not contribute to the selective stabilization of one conformer (90b) over the other (91b). Rather, they act as a chiral probe by increasing the sensitivity of the observed rotational strength associated with any shift in the equilibrium position by a factor of no less than 40! The reason for this enormous increase is that we are now dealing with the very large negative (90b) or equally large positive (91b) contribution of the axial methyl substituent. By means of this technique, the enthalpy

At an IUPAC meeting in 1970 in Riga, Latvia, Günther Snatzke (University of Bochum), who had also been active in variable-temperature circular dichroism studies, is comparing beard length and strength with me.

difference in favor of axial deuterium (91b) over its equatorial conformer 91a was shown[111] to be −9.5 cal/mol. Similar conclusions about deuterium preferentially occupying the position of larger strain were reached in several other instances, which are covered in a detailed review.[112] These observations are entirely consistent with the view that deuterium is of smaller size on the basis of its smaller vibrational amplitude.

The syntheses of the many chiral substrates for these isotope studies are far from trivial, and the interested reader should consult the original literature or construct such approaches on paper. For instance, how does one synthesize 2,2-dimethylcyclohexanone 92 of known optical purity and absolute configuration—a substance whose chirality is entirely due to the fact that one of the methyl groups is based on ^{13}C? Variable-temperature CD studies[108] of that ketone and its trideuterated analog 93 showed that, in these instances, the conformational equilibria are shifted to a slight extent (−1.5 and −3.4 cal/mol, respectively) toward the conformer with the isotope-containing labeled methyl group in the equatorial position.

92: R=$^{13}CH_3$
93: R=CD_3

This account does not mention some of the important concurrent studies performed in other laboratories, notably those of my former student, David Lightner, at the University of Nevada or of Hans Wynberg at Groningen in the Netherlands. Of particular relevance are their syntheses of conformationally immobile systems, such as the two deuterated adamantanones 94 and 95[113,114] and the ^{13}C-labeled adamantanone 96,[115] as well as theoretical studies of Lightner and Bouman.[116] Their work and the numerous other studies performed at Stanford in

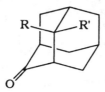
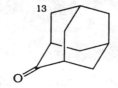

94: R=D; R'=H
95: R=H; R'=D

96

this field are all covered in our 1981 review,[112] which coincidentally was my 1000th publication and essentially also the swan song of 30 years of research on ORD and CD.

Organic Mass Spectrometry

The second new research area, initiated shortly after my arrival at Stanford University, was organic mass spectrometry. In connection with our structural work on cactus triterpenes[3] and alkaloids[4], we also encountered in these plants[117] certain new sterols such as the first naturally occurring 4-monomethyl- and 14-monomethylcholesterol derivatives. These compounds were of considerable biosynthetic significance, because they pointed to the existence of truncated or alternative biological demethylation sequences from lanosterol. The first of these novel cactus sterols was lophenol (97). A key aspect of its structure elucidation[118] was the determination in 1957 of its precise empirical formula, which was accomplished by mass spectrometry in Reed's laboratory in Glasgow. De Mayo and Reed's paper on mass spectral molecular weight measurements of steroid hydrocarbons[119] and Ryhage and Stenhagen's mass spectral studies of long-chain fatty alcohols and acids[120] first drew my interest to this technique, but it was the elegant rationalization by Biemann and collaborators[121] at MIT of the mass spectral fragmentation behavior of alkaloids of the aspidospermine (98) class that stimulated a serious effort at Stanford on organic chemical applications of mass spectrometry.

97

98

It seemed very worthwhile to determine whether the mass spectrometric behavior of steroid ketones—the same substances that opened the entire ORD and CD area in organic chemistry—could be of structural significance. I proposed this line of research to a new Austrian postdoctoral fellow, Herbert Budzikiewicz (now a professor at the University of Cologne), who joined my laboratory in December 1960.

With Lord Todd (left) at the IUPAC Natural Products Meeting in Sydney, Australia, 1960.

Left to right: Australian chemist K. E. Murray demonstrating instrumentation to Sir Robert Robinson, Carl Djerassi, and Roger Adams in Sydney, Australia, 1960.

Our first paper, submitted 8 months later and published in 1962,[122] led us to a significant conclusion:

> ...the mass spectrometric fragmentation patterns of steroidal monoketones show that many structural conclusions are possible, which can locate or at least narrow down the possible points of attachment of a carbonyl group. Indeed, when mass spectrometry is combined with optical rotatory dispersion measurements, then a firm decision can be made in virtually every instance with a total amount of less than 1 mg of substance.

In spite of this optimistic prognosis, I would never have dreamed that 25 years later we would have published 270 papers in a series entitled "Mass Spectrometry in Structural and Stereochemical Problems." Obviously, only a small fraction of this work can be mentioned here, but there is no doubt that our interest in steroids prompted such rapid progress. Rather than proceeding in a chronological sequence, it will be most economical to refer to a series of review articles—most of them from my plenary lectures at various International Union of Pure and Applied Chemistry (IUPAC) conferences—and to four books[123–126] written in collaboration with two of my most productive postdoctorate fellows, Herbert Budzikiewicz and Dudley H. Williams (now a reader at Cambridge University). Since their departure from Stanford in the late 1960s, each of them in his own right has become an internationally recognized authority in organic mass spectrometry.

Our first book[123] had an important pedagogic effect, because in it we employed a simplified mechanistic approach, by assuming that the positive charge, produced when a single electron is removed from the molecule upon electron bombardment in the ion source, is localized at certain centers; that subsequent bond fissions are of the homolytic kind; and that these cleavages can usually be predicted on the basis of standard physical–organic principles. Thus, tertiary carbonium ions or radicals will be energetically favored over secondary and primary ones, allylic or benzylic activation will promote certain bond fissions, and charge localization will be particularly favored in molecules with heteroatoms, where one of the nonbonding electrons is likely to be lost more readily.

Mass Spectrometry in Structure Elucidation

The real emphasis of my first review,[127] entitled "Mass Spectrometric Investigations in the Steroid, Terpenoid and Alkaloid Fields", was on mechanistic rationalizations for structure elucidation. For instance, our studies[128] with estrogens, such as estrone methyl ether (50a), led to the conclusion that two of the most important fragment ions could be visualized in terms of structures 99 and 100. Fragment 99 encompasses rings A and B, together with carbons 11 and 12 of ring C, whereas the ion 100 includes carbon atoms 14, 15, and 16 of ring D. It is thus possible to locate certain substituents in rings C and D solely by consideration of mass shifts in these two fragment ions.

A particularly useful illustration of such mass spectral applications to structural work is provided by our conclusions from the triterpene field.[129,130] All members of the amyrin (101) class of pentacyclic triterpenes undergo electron-impact-induced retro-Diels–Alder (RDA) fragmentation to provide an intense ion (102) retaining rings D and E with part of ring C, which then loses the angular substituent attached to C-17 with generation of ion 103. Members of the taraxerol group (104) also exhibit such RDA processes, except that in this instance, the charge remains with rings A, B, and C (105) because of the different

50a [structure] $\xrightarrow{-e}$

99 [structure] + **100** [structure]

location of the diene system. Such fragmentation of the polycyclic nucleus into two moieties offers an extremely rapid and material-saving approach to localizing substituents in different portions of the pentacyclic system. During the past 20 years, our generalizations have been used by numerous investigators in the structure elucidation of novel triterpenes.

We encountered[131] the operation of a similar RDA process in the *Vinca* alkaloid vincadifformine (**106**), which was the first of numerous indole alkaloids studied in a long-lasting collaboration with Janot and Le Men in France. Retro-Diels–Alder opening of ring C was postulated to produce an isomeric ion **108**, which then underwent fission of the benzylically activated bond with charge retention on the piperidine nitrogen. The resulting ion **109** of mass 124 is by far the most intense one in the entire mass spectrum, and it had been shown by Biemann and collaborators[121] to be characteristic of the aspidospermine (**98**) series, where fragmentation of ring C proceeds with expulsion of ethylene (carbons 3 and 4 in **98**) to an intermediate of type **110**. This process relieves the strain inherent in the fused system around ring C, with concomitant aromatization to an indole nucleus. Ion **110** decomposes further in the manner shown for **108** to produce the diagnostic fragment **109**.

When the double bond in vincadifformine (**106**) was reduced with zinc to dihydrovincadifformine (**107**), a similar Biemann-type[121] fragmentation of ring C was encountered, except that in this instance the elements of methyl acrylate were eliminated to yield **110**, which

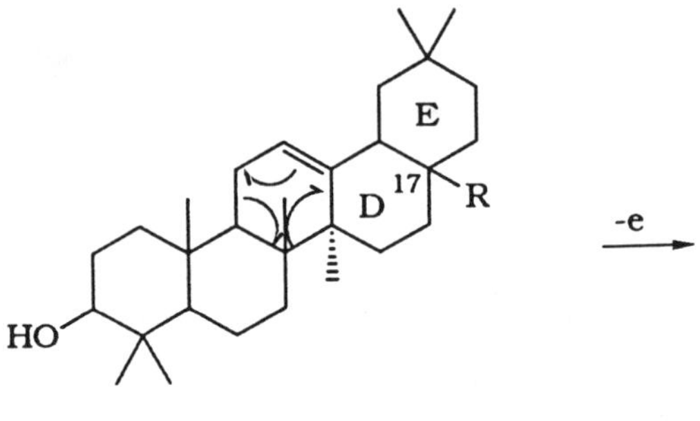

101

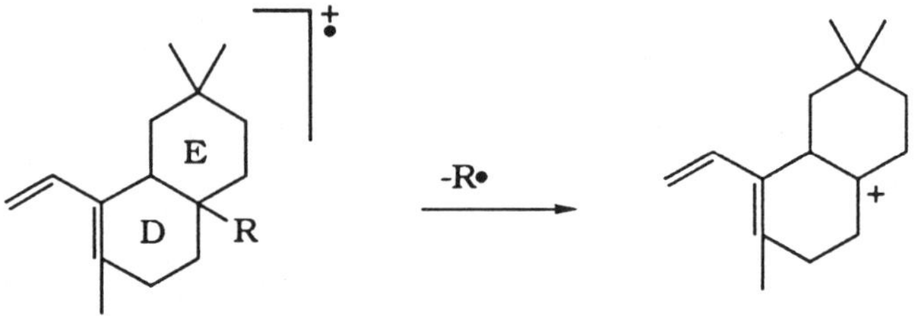

102 R = CH$_3$ or CO$_2$CH$_3$

103

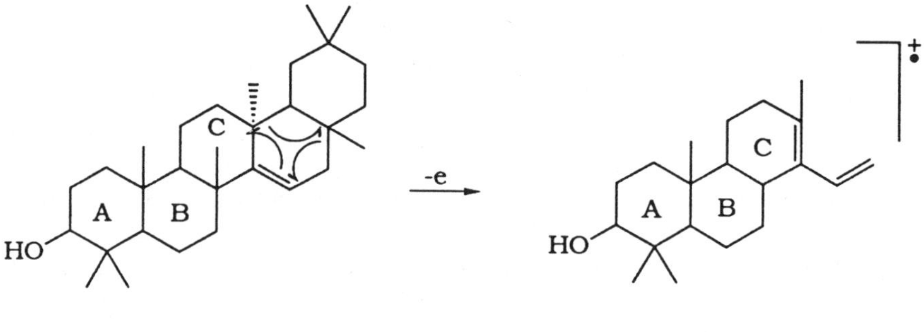

104

105

then led to the intense ion 109. Both vincadifformine (106) and its dihydro derivative (107) gave the characteristic[121] aspidospermine (98) fragment ion 109, whereas only dihydrovincadifformine (107) showed also ring C fragmentation to 110 with loss of carbons 3 and 4 in the form of methyl acrylate. This observation was the chief reason that we assigned[131] structure 106 to vincadifformine and then confirmed it by further chemical transformations.

Many indole alkaloids related to vincadifformine have since been isolated, but it should be remembered that vincadifformine together with its 6,7-dehydro analog tabersonine (111)—whose structure we elucidated[132] through the 2-mass-unit shift in ions corresponding to structures 106–110—have played a key role in indole alkaloid biosynthesis.

111

Their rapid structure elucidation with minute amounts of material has been one of the early triumphs of mass-spectrometry-aided structure elucidation. The publication of an entire book[124] dealing with such applications in the alkaloid field, notably among indole and dihydroindole alkaloids, within a couple of years of the appearance of mass spectrometers in organic chemical laboratories is testimony for the explosive growth of mass spectral techniques.

The importance of the electron-impact-induced retro-Diels–Alder reaction led one of my Stanford faculty colleagues, John I. Brauman, to collaborate with us on a wide-ranging examination[133] of this process in order to determine when the charge is preferentially located on the diene and when on the ene fragment. Our conclusion that this reaction proceeds stepwise, rather than in a concerted fashion, was initially substantiated by our observation[134] that the RDA process among bicyclic octalins (112) was not affected by the stereochemistry of the ring juncture. Subsequent studies[135] with pairs (5α and 5β stereochemistry) of Δ^7-steroid olefins (113) however, did demonstrate marked stereochemical dependence consistent with a concerted quasi-thermal mechanism. We concluded[135] that both mechanisms can operate after electron impact, the key factor being whether the stereochemistry of the ring juncture is retained after electron impact.

112 R = H or CH$_3$

113

Isotope-Labeling Studies

In attempting to elucidate the mechanisms of mass spectral fragmentation processes, one must be in a position to determine which atoms of a given molecule are retained in the charged fragment and which are lost in the neutral moiety. Frequently, one can reach such conclusions by simple consideration of the mass of the charged fragment, although in the presence of certain heteroatoms, high-resolution mass spectral measurements must be performed to distinguish between isobaric fragments (e.g., N vs. CH_2). There are also two chemical ways of approaching the problem. The easier one is by substituent labeling, for instance, by examination of a homolog with a methyl group. If the latter does not affect the course of the particular fragmentation, then the presence or absence of the appropriate mass shift will pinpoint the portion of the molecule retained in the fragment.

By far more precise and informative is the substitution of one of the constituent atoms by an isotope—most frequently ^2H or ^{13}C and, less frequently, ^{15}N or ^{18}O. At a very early stage in our work, I decided

At the 1964 IUPAC meeting in Kyoto, Japan, I reviewed our work on structure elucidation by means of mass spectrometry. Here I am greeting an old Japanese friend, Ken'ichi Takeda (Shionogi). Vlado Prelog is sitting at the end of the same row. As usual, I have my stiff left leg propped up on the portable stool that I always carry.

that we would make the necessary investment of time and manpower in extensive isotope labeling. In retrospect, this turned out to be the most important strategic decision that could have been made in this field.

For reasons that will become obvious in the sequel, we focused on deuterium labeling. In addition to many conventional methods of introducing deuterium, which we summarized on several occasions,[136,137] we also developed several new ones that occasionally carried the additional bonus of shedding mechanistic insight on various synthetic reactions that so far had only been performed with proton-containing reagents. Some examples are the effect of solvent polarity on the course of homogeneous catalytic hydrogenation of olefins and unsaturated ketones,[138] and the mechanisms of the reduction of tosylhydrazones with lithium aluminum or sodium borohydride,[139] of α,β-unsaturated ketones with diborane,[140] of α,β-unsaturated tosylhydrazones with sodium cyanoborohydride,[141] and of 1-alkyn-3-ols with lithium aluminum hydride.[142] Except for the last example, the course of all other reactions was demonstrated in the steroid series.

Because of space limitations, rather than a lack of appreciation, I cannot even allude to the numerous multistep syntheses performed by many talented collaborators that provided the deuterated steroids and

Left to right: Ken Rinehart, Al Burlingame, and Fred McLafferty—three important organic mass spectrometrists—in April 1965, with a CEC 21–110 mass spectrometer at Purdue University.

other labeled molecules. I can only describe some of the highlights[143–146] of the mass spectral mechanistic deductions that could be reached with such labeled substrates.

Mass Spectra of Steroid Ketones and the McLafferty Rearrangement

In an attempt to determine how effectively a carbonyl group acts as a charge localizer and, hence, as a director of subsequent mass spectral fragmentations, we undertook a systematic study of the mass spectral behavior of all 11 possible nuclearly substituted steroid monoketones—each of them with specific deuterium labels in relevant positions. Aside from significant insight into the manner in which such monoketones fragment, an important mechanistic conclusion could be reached concerning one of the most widely studied mass spectral fragmentations: the McLafferty rearrangement,[147] involving β-bond fission in carbonyl compounds with transfer of the γ-hydrogen to the oxygen atom. Most studies of the McLafferty rearrangement had been conducted with aliphatic ketones or esters (114), which present no major barrier to close approach of the departing hydrogen to the receptor oxygen in a six-membered cyclic transition state. In cyclic systems the situation is different, and we felt that studies with appropriately labeled steroid ketones would shed some light on the interatomic distance requirements in the McLafferty rearrangement.

114

Four relevant examples are collected in Table I. The 16-ketocholestane showed[148] clean transfer of the C-22 hydrogen, whereas the 11-ketoandrostane labeled with deuterium at positions 1[149] or 19[150] failed to undergo such hydrogen migration. The question can be asked whether this apparent anomaly in 11-keto steroids is sui generis or of a more general nature. In the 16-keto steroid, the C-22 hydrogen atom can approach the oxygen to within 1.5 Å, whereas in the 11-ketone, this distance increases to 1.8–2.2 Å, depending on whether the C-1β or C-19 hydrogen atom is considered. To settle this question, a 7β-deuterio-15-ketone was synthesized,[151] in which the interatomic distance is 2.3 Å (Table I). In this substance, the deuterium atom was not transferred to oxygen. Our results thus demonstrated for the first time that the hydrogen–oxygen distance plays a very important role and that the limit for such six-membered concerted collapse is somewhere between 1.5 and 1.8 Å.

Table I. Nuclearly Substituted Steroid Monoketones

Structure	H–O Distance (Å)
(structure with positions 22, 16, H, O)	1.5
(structure with positions 1, 11, H, O)	1.8
(structure with positions 19, 11, H, O)	2.2
(structure with position 15, H, O)	2.3

Mass Spectral Fragmentation of α,β-Unsaturated Steroid Ketones

A further illustration of the unique advantage of deuterium labeling can be provided from our extensive studies[152,153] of the mass spectral fragmentation of α,β-unsaturated-3-keto steroids such as 5 (typical of the sex and adrenal hormones), 115 (present in most synthetic corticosteroids), and 116. The most diagnostic feature of their mass spectra is

double fission of ring B, as indicated schematically by the wavy lines in 5, 115, and 116. In the first two instances, simple arithmetic shows that generation of ions with masses of 124 and 122 from 5 and 115, respectively, must be accompanied by transfer of two hydrogen atoms from the neutral portion of the molecule. By identifying the origin of these migrating hydrogens (i.e., C-8 and C-11 in the case of 115) through deuterium labeling,[152,153] we could draw a rather precise picture for the course of this complicated electron-impact-induced fragmentation.

The corresponding Δ^1-3-keto steroid 116 seemingly behaved differently, because the ion of mass 122 (see wavy line in 116) does not require (arithmetically speaking) any hydrogen migration during rupture of ring B. However, deuterium labeling[152] disclosed the existence of a *reciprocal* hydrogen migration, with hydrogens attached to C-5 and C-8 playing the key role. With this information, which could only have been derived from deuterium labeling, one could draw a plausible reaction path via intermediates 116a and 116b.

Charge Localization in Ethylene Ketals

One of the most dramatic instances of charge localization involves steroidal ketals.[154] Saturated 3-keto steroids display[122] a complex mass spectrum with numerous peaks, whereas the corresponding ethylene ketal 117 has virtually only two fragment peaks (m/z 99 and 125, depicted by wavy lines in structure 117) in addition to the molecular ion.[155–157] A very plausible fragmentation mechanism could be depicted through deuterium labeling,[156,157] based on the assumption of charge localization on the ketal oxygen functions followed by α fission to give rise to either species a or c. Hydrogen migration through a six-membered transition state is favored, in that the primary radical site (C-4 in a; C-2 in c) is replaced by an allylically stabilized radical ion (b or d). In the case of d, simple homolysis of the C-1—C-10 bond leads to the most intense ion (m/z 99), which can be depicted by structure 118. In d, there are two choices for homolytic fission, with the more highly substituted C-5—C-10 bond being preferred. A second six-membered transition hydrogen transfer—this time from C-6 (e)—followed by cleavage of the C-7—C-8 bond in f offers a perfect rationale for the generation of the second most intense fragment ion 119 (m/z 125).

The electron-impact-induced rupture of the ethylene ketal 117 is of obvious structural utility. For instance, if the parent 3-ketone were monoalkylated and one wished to establish the site of α methylation (C-2 or C-4), conversion to its ethylene ketal and determination of its mass spectrum would settle the question unambiguously. If the methyl group is at C-2, the m/z 99 peak (118) would then be observed at m/z

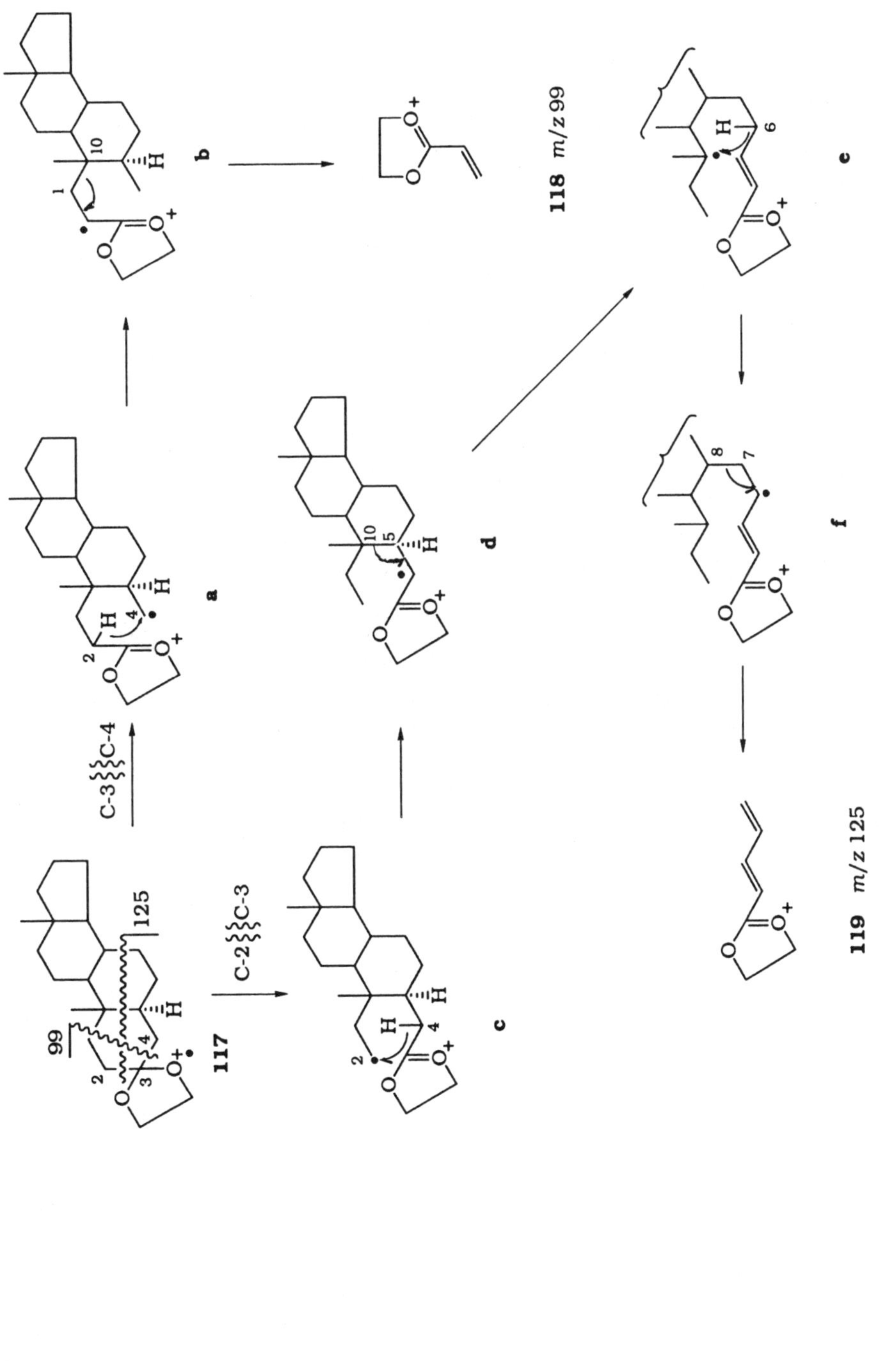

DJERASSI *Steroids Made It Possible*

118 m/z 99

119 m/z 125

113; conversely, if the methyl group is attached to C-4, the m/z 125 peak (119) will be shifted to m/z 139.

Mass Spectra of Saturated Steroid Hydrocarbons

Even though it was the most time-consuming and complicated of our steroid mass spectrometric studies, our deuterium-labeling work in the steroid hydrocarbon field[158,159] was directly responsible for our subsequent research on marine natural products.

The concept of preferred charge localization is least likely to succeed in saturated hydrocarbons. Indeed, it cannot be employed in the interpretation of mass spectra of saturated straight-chain hydrocarbons, such as n-docosane (Figure 2).[160] All fragment ions are of nearly identical intensity, demonstrating that most bond fissions occur with equal facility. The situation is different with polycyclic hydrocarbons, such as the steroid cholestane (Figure 2), because different bond cleavages can give rise to tertiary, secondary, and primary carbonium ions and also relieve steric strain to a different extent. The two most ubiquitous and also most frequently studied steroid fragmentations are the ones that give rise to the m/z 217 and 218 peaks (cf. Figure 2).

Reed[161] suggested the process indicated in 120, for which a double hydrogen transfer had to be postulated. Friedland et al.[162] revised this proposal to that shown in 121—the presumed virtue of the revised picture being that it involved no hydrogen transfer. Finally, Ryhage and Stenhagen[163] advanced formulation 122, which requires a single hydrogen migration. In addition to the different hydrogen migrations, different structural conclusions would be derived from each of these proposals. Both 120 and 121 demand the loss of the C-18 angular methyl group; in 122, carbon atoms 15 and 16 are lost with the side chain, whereas in 121 C-15 is retained, and in 120 both C-15 and C-16 are retained.

120   **121**   **122**

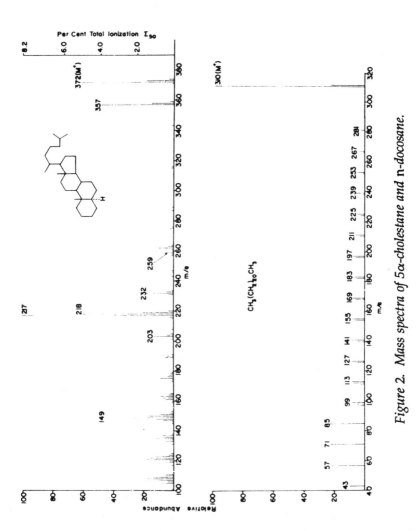

Figure 2. Mass spectra of 5α-cholestane and n-docosane.

Our first deuterium-labeling studies[164] with steroid hydrocarbons demonstrated unambiguously that the Swedish[163] proposal (122) was correct. The origin of the migrating hydrogen atom, however, turned out to be difficult to identify because of multiple sites (*see* percentages indicated in structure 122) that required heroic labeling efforts by many of my collaborators. The least expected position, C-14, was found to constitute the most important source (75%), but it was also the one that was most sensitive to stereochemical variations in the C–D and B–C ring junctures.[165] These subtleties are of mechanistic significance, but fortunately, they do not detract from the structural utility of the important D-ring fragmentation implicit in 122, which proved to be of crucial importance for the structure elucidation of dozens of novel marine sterols (vide infra).

The second peak associated with ring-D rupture occurs at m/z 218 (Figure 2) and becomes increasingly more important as the electron voltage is lowered. Originally,[166] the ion of mass 218 was represented as simple ring-D fission without associated hydrogen transfer (123→124), which led to the conclusion that the expelled neutral species, encompassing the side chain with carbon atoms 15, 16, and 17, was the cyclopropane 125. However, our deuterium labeling[164,167] demonstrated that the situation was both more complicated and also more rational in terms of organic chemical stability concepts.

The molecular ion species responsible for the ring-D fission leading to both the m/z 217 and 218 peaks (Figure 2) can be visualized in terms of 123. Not only does 123 display the best radical–ion characteristics (tertiary carbonium ion; secondary radical) that can be generated from a steroid nucleus, but it also lacks the steric strain of the *trans*-hydrindane system. Instead of undergoing direct rupture of the C-14–C-15 bond in 123 to the ionized olefin 124 (m/z 218) and the neutral cyclopropane 125, the C-18 proton migrates to the radical site at C-17 to provide the ionized olefin 123a. Back-migration of the C-16 proton and homolysis of the C-14–C-15 bond then produces the ion 124a (m/z 218) and the neutral olefin 126. The greater stability of the olefin (126) relative to the cyclopropane (125) must provide the driving force for this reciprocal hydrogen transfer. In terms of quasi-equilibrium theory, the lower activation energy of the hydrogen rearrangement process outweighs the higher frequency factor of the simple cleavage (123→124 + 125).

Mass Spectra of Unsaturated Steroid Hydrocarbons

By examining a series of nuclearly substituted olefins (127),[145,168] we were able to draw mechanistic conclusions concerning the effect of nuclear unsaturation on side-chain loss and ring-D fission. Deuterium labeling, associated with steroids containing unsaturation in the side

127

chain,[146,169,170] proved to be particularly useful. A typical illustration is offered by the sponge sterol stelliferasterol (128), which was one of the first C_{30} sterols to be uncovered in nature.[171] Four diagnostic mass spectral peaks, indicated schematically in 128, practically defined the structure of this novel marine sterol. The ring-D fission peak at m/z 231

128

and the peak at *m/z* 273 associated with side-chain loss demonstrated that all three extra carbons beyond the conventional C_{27} empirical formula of cholesterol must reside in the side chain. The two side-chain fragmentation peaks at *m/z* 328 and 314 had been shown in earlier model compounds[170] to indicate two alternative McLafferty rearrangements (128a→129 and 128b→130) typical of a C-25–C-26 double bond. All that was needed for a complete structure elucidation was NMR evidence for the presence of an ethyl substituent and two methyl groups on the double bond.

128a   **129** *m/z* 328

128b   **130** *m/z* 314

Electron-Impact-Induced Functional-Group Rearrangements

Our isotopic-labeling studies had other mechanistic ramifications beyond what could be deduced from the course of hydrogen transfers. For instance, we were the first to demonstrate through multiple labeling the occurrence of electron-impact-induced 1,2-alkyl and aryl rearrangements[172,173] in unsaturated ketones. Some indication of such a process was given by the shifts of the m/z 69 peak in various labeled analogs (see asterisks) of androst-1-en-3-one (131).[152] Because the spectrum of 131 is very complicated, we sought to simplify the situation by examining the bicyclic octalone 132.[173] Its mass spectrum displayed an intense ion of mass 69, to which structure 133 could be assigned unambiguously on the basis of labeling studies (see asterisks in 132) and high-resolution mass spectral measurements. These results prompted us to examine the ultimate structural limit for such a skeletal rearrangement; 4,4-dimethylcyclohex-2-en-1-one (134, R = H) did not show such a peak, but the 4,4,5-trimethyl analog (134, R = Me) did. This in turn opened the way to the first measurement[173] of electron-impact-promoted relative migratory aptitudes. The mixed aryl–alkyl-substituted cyclohexenone 135 displayed a 10-fold preference for phenyl migration (135→136) compared with methyl migration (135→133).

Another type of rearrangement, which was uncovered with Mark Green (now a professor of chemistry at Polytechnic Institute of New York), involved the interaction of remote functional groups after ionization in the mass spectrometer. The first example was 4-methoxycyclohexanone (137), whose mass spectral base peak was shown[175] by high-resolution mass spectral measurements to occur at m/z 74 and to correspond to the elemental composition $C_3H_6O_2$ (139). Because the two oxygen atoms in the parent molecule 137 were four carbon atoms apart, a rearrangement must have occurred to produce an ion containing three carbons and two oxygens. Detailed studies[175] involving a variety of model compounds, combined with deuterium labeling, delineated the scope of this functional-group rearrangement. The process, which occurs with equal facility in 4-hydroxycyclohexanone, is best described through the intermediacy of ionized methyl 5-hexenoate (138), which in turn, is generated from the α-cleavage ion 137a by attack of the electron-rich methoxyl oxygen on the electron-deficient carbonyl carbon atom, followed by hydrogen migration and ring opening (137b).

A knowledge of the operation of such skeletal rearrangements is absolutely crucial if the mass spectral technique of element mapping[176] (Biemann's approach to defining the minimum carbon distance between heteroatoms, based on high-resolution mass spectral measurements) is to be considered a reliable tool for structure elucidation. For this reason, the late Peter Brown (professor of chemistry at Arizona State University

98   PROFILES, PATHWAYS, AND DREAMS

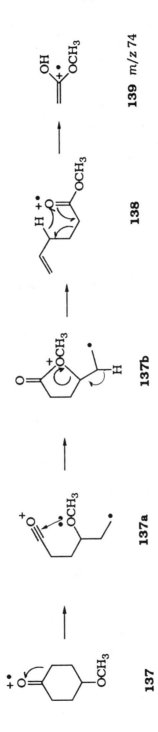

after his departure from my laboratory) and I summarized this field in an extensive review[177] with special reference to electron-impact-induced rearrangements of atoms other than hydrogen.

The success of our steroid studies and our conviction that application of simple organic chemical mechanistic principles are of enormous aid in interpreting mass spectral fragmentations led us to a much broader research effort in mass spectrometry: an examination of virtually all common organic chemical functionalities. Our results, as well as those of many other investigators, were summarized in a 690-page book;[126] it turned out to be one of my most-cited scientific contributions.[178]

Magnetic Circular Dichroism

Because the leitmotif of this book is my intellectual infatuation with steroids, what is the rationale for including here a brief discussion of the third major new area of research, organic chemical applications of magnetic circular dichroism, which was started at Stanford? There are two reasons: first, because the original impetus arose directly from our earlier ORD and CD studies; and second, because the carbonyl chromophore—so important in our steroid chiroptical work—entered again, and this time in an unexpected manner.

The principal limitation of ORD and CD is the requirement that the experimental substrates be optically active. In theory, this handicap might be overcome by generating magnetically induced rotation (Faraday effect), for which no asymmetric molecules are required. Our entry into this field was stimulated by the appearance in the early 1960s of publications by Shashoua[179] and Briat[180] on the magnetic optical rotation (MOR) of certain paramagnetic and diamagnetic substances, and the important review by Buckingham and Stephens[181] on theoretical aspects of magnetic optical activity. For instrumental and practical reasons, Edward Bunnenberg (who collaborated with me in this area until his death in 1984) and I chose magnetic circular dichroism (MCD), rather than MOR, as the preferred technique. We persuaded David Schooley (now a professor of biochemistry at the University of Nevada) to focus in his Ph.D. thesis upon exploratory MCD measurements of certain aromatic organic molecules.[182] Shortly thereafter, Bernard Briat, who had carried out some of the early MOR work in Paris,[180] came to Stanford for a year of postdoctorate research. He participated in our first porphyrin studies with optically active chlorins[183] in order to effect a direct comparison of CD with MCD, because of the fundamental difference between natural and magnetically induced optical activity. In natural optical activity, the disturbing factor is internal to the molecule, and its own dissymmetry is responsible for the fact that upon excitation the electrons follow a helical path. The chirality of the helix is responsible for the differential absorption of right and left circularly polarized light. In magnetically induced optical activity, the disturbing element—the magnetic field—is external so that the molecule's geometry should not play an important role.

In spite of our assumption[184] that the magneto–optical behavior of achiral carbonyl compounds would not be influenced by conformational factors or the structural environment (an opinion that was strengthened by the reported[179,185] lack of success in securing satisfactory MOR or MCD spectra of optically inactive ketones and aldehydes), Günter Barth and several other members of my research group undertook such a study anyway. Barth had just joined my group from Berlin and eventually participated in our CD and MCD studies for over a dozen years. Our experimental success[186] was due to a large extent to our utilization of a superconducting magnet and a circular dichroism instrument of increased sensitivity, which had been donated by JASCO, a Japanese instrument manufacturer. In this manner, we could easily secure satisfactory MCD spectra for all achiral carbonyl compounds. The MCD spectra showed a variety of shapes and amplitudes, which indicated that the technique was much more sensitive to the symmetry (i.e., structural environment) around the carbonyl group than had been envisaged originally.

These surprising results justified a substantial theoretical effort in this field, which was first started in a renewed collaboration[187] with Albert Moscowitz of the University of Minnesota and then performed in-house[188] when Robert Linder joined my group. Over a period of approximately 20 years, with nearly 70 research publications, my Stanford group surveyed the MCD behavior of a variety of chromophores.[189] We also developed certain special analytical applications of MCD, such as a very sensitive quantitative determination of tryptophan in intact proteins.[190] However, our main focus remained on carbonyl compounds and porphyrins, with the most dramatic practical and theoretical advances occurring in porphyrins.

In our own laboratory, much of the experimental work on the MCD spectra of biochemically important, optically active molecules was performed by John Dawson, now professor of chemistry at the University of South Carolina, as part of his Ph.D. thesis and by J. R. Trudell, who had received his Ph.D. with me in 1969 and in the interval had become an associate professor in the Department of Anesthesia in our medical school. This research included various cytochromes,[191] evidence for thiolate ligation,[192] and a comparison of chloroperoxidase with cytochrome P-450.[193] However, starting with a systematic examination of the MCD spectra of porphyrin dications,[194] dianions,[195] and metal complexes,[196] Barth, Linder, and Bunnenberg of our group concentrated on optically inactive porphyrins because of the wealth of spectroscopic[197] and theoretical[198] information that could be extracted from them by means of MCD.[199] This work resulted in a demonstration by MCD of N–H tautomerism in certain porphyrins,[200] which was confirmed by NMR studies[201] with ^{15}N-labeled analogs in collaboration with H. Limbach of the University of Freiburg, and most importantly, in a demonstration of the utility of MCD for detecting out-of-plane conformational changes in π-substituted porphyrins.[202] Major theoretical contributions to this latter aspect were made by Robert Goldbeck, who joined my group as resident theoretician after the departure of Robert Linder for the Surface Science Laboratory Company.

Our MCD research depended on a very substantial synthetic effort in the porphyrin field, in which we had never been active before. This effort required the participation of several pre- and postdoctoral co-workers, whose names appear on our MCD publications.[197,198,200,202] In addition to this synthetic work, our involvement in porphyrin chemistry prompted us to reexamine our early mass spectral investigations in this field,[203] which had led to the conclusion that mass spectrometry is not useful for the structural determination of such molecules (e.g., location of discernible substituents on particular pyrrole rings) because of the paucity of fragmentations undergone by the porphyrin nucleus.

These conclusions had been based on conventional electron-impact mass spectra. However, using newer techniques—notably chemical ionization desorption mass spectrometry, with ammonia[204] rather than hydrogen[205] as the reagent gas—we encountered extensive fragmentation of the porphyrin macrocycle, with predominant formation of protonated mono- and dipyrrolic species. Substituents on the pyrrole rings were found to largely retain their structural integrity, because secondary cleavage processes hardly occur under such mild conditions. Sequence information about the pyrrole rings can thus be deduced from the fragment ions in the dipyrrolic region. These results, stimulated by our MCD studies, represent a significant contribution to porphyrin chemistry that is way beyond the narrow confines of magnetic circular dichroism. It is fair to say that we would never have entered the porphyrin field had it not been for our work on steroid ketones.

Applications of Computer Artificial Intelligence Techniques

During Stanford University's initial courting of me in early 1959, three faculty members, aside from the legendary provost F. E. Terman, spent much time with me: Joshua Lederberg, then chairman of the Genetics

Opening of the Syntex Institute of Molecular Biology in the Stanford Industrial Park, circa 1962. Left to right: Carl Djerassi, Charles Allen, Jr. (investment banker from Allen & Company, New York, NY), and Joshua Lederberg.

Department and since 1978 president of Rockefeller University; Arthur Kornberg, chairman of the Biochemistry Department; and the late Henry Kaplan, then chairman of the Radiology Department. All three became personal friends, but the longest and professionally most intimate relationship was with Joshua Lederberg. After I had accepted the Stanford offer in 1959, but while I was still living in Mexico City and commuting periodically to the Bay area, Lederberg and his wife visited us in Mexico. They even introduced us to the architect who built our California home in time for our September 1960 arrival.

Within 2 years, Lederberg was serving as principal scientific advisor of the Syntex Institute for Molecular Biology, which I had induced my fellow Syntex directors to establish on the Stanford Industrial Park and for which I took overall responsibility. I had felt that Syntex should diversify beyond steroids, and molecular biology seemed an ideal vehicle: It promised exciting new science, potential medical applications, and, most importantly, the presence of several academic pioneers on the Stanford faculty. The Syntex Institute, though small, was probably the earliest American industrial laboratory dedicated specifically to exploring practical applications of the burgeoning field of molecular biology.

Fred Terman, Stanford's provost and virtual founder of the Industrial Park, was delighted that a chemically and biomedically oriented corporation was prepared to join what, until then, was a group of corporations exclusively dedicated to electronics, computers, and publishing. Within a year, I persuaded my fellow board members to establish Syntex's United States operations in Palo Alto. By the middle 1960s most of Syntex's research had also moved from Mexico City to Palo Alto to join the Molecular Biology Institute, which was the beginning of the company's research efforts outside the area of steroid chemistry and medicine. Twenty years later, Syntex, by now a billion-dollar company, has become the largest occupant of the Stanford Industrial Park and has spawned several other research-oriented companies such as ALZA, SYVA, and Zoecon. The middle 1960s was also the period when my polygamous professional life really flourished. When my close friend and colleague, Alejandro Zaffaroni, left his position as president of Syntex Research to found ALZA and pursue his interest in novel drug delivery methods, I succeeded him as president without giving up my Stanford academic position. At the same time, I helped found (and served as board chairman of) SYVA—a joint venture of Syntex and Varian—which became one of the truly innovative companies in the medical diagnostic field. Zoecon, which was created in 1968 to concentrate on novel approaches to insect control, became my real industrial baby. In 1972, I discontinued my connection with Syntex, and thereafter focused the industrial portion of my bigamous life on Zoecon as its CEO until it was acquired in 1983 by the Swiss pharmaceutical company Sandoz.

Aside from our social interaction, I saw Lederberg professionally several times per week, usually at lunch. Until today, I have never met anybody with such a quick mind and such broad interests. There was no topic—not even esoteric chemical ones—that could not be discussed with him and that would not be countered with insightful comments. It was not surprising, therefore, that Lederberg became quite familiar with our flourishing mass spectrometric research during a period when he, the Nobel-Prize-winning bacterial geneticist, had become deeply interested in exobiology.

Lederberg was one of America's chief proponents of unmanned flights to outer space and had established, with the assistance of Elliot Levinthal (who also became a good friend), a sophisticated instrumenta-

One of the most enjoyable events during my Zoecon career was the visit of King Carl XVI Gustaf of Sweden with a group of Swedish industrialists. Here he is looking amusedly at some cockroaches held by Gerardus Staal (director of insect research at Zoecon). Between the king and me is Karl-Erik Sahlberg (president of Perstorp AB), and at the right is Bengt Modeer of the Swedish Academy of Engineering Sciences.

tion laboratory for the development of devices that might detect evidence of life in the forthcoming lunar and Mars missions. Mass spectrometry was high on his list of instrumental priorities and, because he was focusing on unmanned flights, computer control of the various instruments and experiments was crucial. Lederberg's interest in automated instrumentation for exobiology led him to apply topological graph theory to the classification of organic molecules, the forerunner of DENDRAL[206] (acronym for dendritic algorithm). He subsequently collaborated with Edward Feigenbaum on the implementation of DENDRAL, in what came to be known as knowledge-based systems of artificial intelligence. Feigenbaum was one of the founders of the field of computer artificial intelligence and subsequently became the chairman of Stanford's Department of Computer Science. Their collaboration lasted for over 15 years and, with many research fellows and students, produced the famous DENDRAL system,[207] which formed the basis of the computer-aided structure elucidation approaches in which I participated.

With Ed Feigenbaum at his 50th birthday party in 1986.

Applications in Mass Spectrometry

Lederberg's involvement in mass spectrometry started in 1963, when he developed a very convenient algorithm for calculating molecular formulas from high-resolution mass spectral data. With the advent of high-resolution mass spectrometers and penetration of ever-higher mass ranges, questions such as "given a molecular ion peak of 718.3743 ± .0060, what empirical formulas are consistent with such a value?" are hardly trivial. Lederberg's clever and convenient solution was published in abbreviated form as an appendix to one of our books[125] and subsequently in an independent volume.

Insofar as mass spectrometry was concerned, Lederberg's exobiology problem could be stated in simple terms: What is the maximum structural information that can be extracted from a mass spectrum measured on Mars and returned to earth by telemetry? In this context, Lederberg and Feigenbaum invited me in 1967 to join their DENDRAL project as the chemical member of a triumvirate that would focus on "Applications of Artificial Intelligence for Chemical Inference." This became the fourth new research area undertaken by my research group at Stanford and also the lead title in a series of publications that culminated in my 1982 IUPAC lecture, "Computational Aids to Natural Products Structure Elucidation."[208]

The first group of studies laid the groundwork for computer-aided interpretation of low-resolution mass spectra.[209] Key contributors from my own group were an Australian research associate, Alan Duffield (now a professor at the University of New South Wales), who had taken Herbert Budzikiewicz's position as manager of my mass spectrometry laboratory; Alexander Robertson, on sabbatical leave from the University of Sydney; Gustav Schroll, from the University of Copenhagen; and Armand Buchs, professor and later chairman of the Chemistry Department of the University of Geneva.

Before generating a feasible number of candidate structures from a given mass spectrum, it is necessary to calculate the number of possible structures for a given empirical formula. The drastic increase in the number of such structures is shown by the fact that, whereas only four isomers are possible for C_3H_9N, there exist 14,715,813 isomers for $C_{20}H_{43}N$. Therefore, our very first joint paper[210] with the Lederberg–Feigenbaum group (which from the very beginning included Bruce Buchanan, now professor of computer science at the University of Pittsburgh) dealt with the use of DENDRAL for solving this structure-generation problem among acyclic structures. Soon thereafter, we were able to extend this approach to cyclic structures.[211]

Chemical structures containing either chemical absurdities or undesired functional groups (e.g., unfavored tautomers) were eliminated by a BADLIST command. Our first successful demonstration of the then-unique capability of DENDRAL to manipulate structural representations of organic molecules and their functional groups and to generate rigorously exhaustive and irredundant lists of candidate structures was performed with my favorite chromophore—specifically, with aliphatic ketones.[212]

The procedure for computer interpretation of an unknown ketone proceeded as follows: The program is given a complete low-resolution mass spectrum and the empirical formula of the substance. By using a theory of mass spectral fragmentations, which essentially represents the appropriate chapter for a given functionality from our book,[126] the PRELIMINARY INFERENCE MAKER decides what functional groups could be present in the unknown compound. The inferred functional group gets the highest priority by the program and is placed on GOODLIST, whereas rejected variants are segregated into BADLIST and not considered further. The STRUCTURE GENERATOR, using the DENDRAL algorithm, then constructs all possible candidates within the constraints inferred from the mass spectrum. Each of these alternatives is scrutinized by the PREDICTOR, which selects the most significant peaks of a hypothetical mass spectrum (constructed from the guidelines contained in our book[126]) for each candidate structure. The program accepts or rejects any structure on the basis of such comparison between known and predicted mass spectra and finally arranges the admissible structures in an order generated by a SCORING FUNCTION.

DENDRAL was tested with equal success in the interpretation of the mass spectra of aliphatic ethers[213] and amines.[214] With ethers, using solely mass spectral inputs, our program either offered one or two candidates (if the number of theoretical possibilities is less than 200) or resulted in a drastic reduction of the number of possible structures (e.g., 10 out of 989 possibilities). If NMR data were also employed, then further reduction—usually leading to a single (correct) structure—was possible.

In 1971, Dennis Smith joined my research group as a postdoctorate fellow. His arrival coincided with a major escalation in the scope and sophistication of our artificial-intelligence efforts. Because DENDRAL uses a theory of mass spectrometry much as the mass spectroscopist does, the chemist is in a position to understand the reasoning

The annual cross-country skiing trips with my research group included occasional chemical discussions in the snow. Dennis Smith, one of the key members of our computer group, is the fifth from left. I am standing in the middle, with particularly long ski poles because of my stiff leg.

steps of the program. Thus one can suggest extensions or alternative steps when the program fails to analyze some spectra correctly. I felt that the time was ripe to jump from alicyclic, monofunctional models into a complex natural products group. Given my personal predilection for steroids, we chose the estrogenic hormones because of their biological importance and because our earlier studies[128] had demonstrated strong correlations between mass spectra and structure. The enormous number of possible elemental compositions based on low-resolution mass spectra of complex molecules is an effective deterrent to the successful implementation of the relatively crude approach that we had developed for simple ketones, ethers, and amines. It was clear that our program had to be modified to encompass the specificity of complete high-resolution mass spectra, where elemental compositions of all ions are determined.

We demonstrated the power of our increasingly more sophisticated approach in several ways.[215] First, the program was offered information about the basic structural unit of a class of compounds (in this instance the estrogen skeleton 140), together with the mass spectral fragmentation mechanisms general to that class, as well as a complete high-resolution mass spectrum and any metastable-ion information. With

140

this input, the program's performance was comparable in quality, but far superior in speed, to the performance of a trained mass spectroscopist in interpreting the mass spectra of over 40 different estrogenic steroids. Next, we showed that by using metastable-ion information, we could analyze complex mixtures of estrogenic hormones without prior separation. Finally, we introduced a program, INTSUM, which provided a means for the systematic interpretation and summary of evidence for all possible mass spectral fragmentations of a set of related molecules (e.g., estrogens).

Computer-Aided Structure Elucidation: CONGEN

At this stage, Ray Carhart, who had just received his Ph.D. with J. D. Roberts at Caltech, arrived in my laboratory and turned his attention to an expansion of the DENDRAL program beyond mass spectrometry to data provided by NMR techniques. We selected ^{13}C NMR,[216] because with a Danish collaborator, Hanne Eggert, we had carried out a systematic study on additivity relationships in ^{13}C NMR chemical shifts among various steroids.[217] Including NMR data was only one step in what eventually became an integrated approach, using a variety of spectral and biosynthetic data such as the isoprene rule, to use artificial intelligence for structure elucidation of organic compounds. Carhart played a key role in the creation of the CONGEN (an acronym for constrained structure generation) program.[218]

CONGEN generates all structural isomers without duplication, consistent with the information (chemical, spectral, biosynthetic, etc.) provided in the form of inferred structural fragments (called superatoms) and a variety of constraints on structural features that are desired (e.g., six-membered ketone, based on infrared data) and undesired (e.g., no five- or four-membered ketones, based on the same infrared data). An important aspect of CONGEN is that it allows the stepwise solution of a structural problem through intermediate partial structures, which can be examined interactively and constrained further during the course of generation of the final solution. This may involve performing certain laboratory experiments to answer questions raised by the computer's generation of unexpected structural alternatives.

CONGEN, though not our final refinement to computer-aided structure elucidation, had enormous pedagogic value. I built around it an entire graduate course on the use of physical methods in structure elucidation because of its extraordinary humbling effect. Because the user, as distinct from the programmer, needs to employ only conventional chemical symbolisms and standard English, it was possible to expose the entire class quickly to the general rules[218] of CONGEN. Each student was asked to select a publication from the recent literature in which the structure of a natural product was reported on the basis of chemical and spectral data, but without the ultimate proof of X-ray analysis or total synthesis. The students had open access to computer terminals and were asked to feed all information from that paper into CONGEN to determine whether the program agrees with the published structure. The experience was indeed a humbling one because, according to CONGEN, in *no instance* was the published structure the only one consistent with the input data. At least one other and frequently over a dozen alternatives were produced, and I used a few examples from this class experience in a review article[219] to illustrate the power and utility of this method.

An amusing sideline of this pedagogic exercise was the response of the "victims" from the literature. I wrote to each author, describing my students' (or really the computer's) conclusions, and asked whether they had any comments about the ambiguity of their published results. The replies fell into three categories.

The majority provided me with the usual "Ah well, but . . ." response and then cited some additional data—usually NMR or other spectral information— that was not contained in the publication, but which the outraged author then dug up in order to demolish one of the computer-generated alternatives. My answer, of course, was that these data should have been in the paper in the first place, and I suggested, only partly tongue-in-cheek, to one of the journal editors that CONGEN could become a completely automated and totally unbiased journal referee for any paper dealing with structure determination. I even included this proposal in print,[219] but to my knowledge no journal so far has had the courage to try the experiment.

The second group never replied—possibly out of shock or chagrin, but the third group was the most pleasing. Their reply was usually couched in terms such as, "How can I get a copy of your program?"

Computer-Aided Structure Elucidation: GENOA

A major improvement on CONGEN was GENOA (an acronym for structure generation with overlapping atoms), which essentially became the final version[220] of our structure-elucidation program. Its most important feature was its ability to handle redundant information, which is usually obtained in the real laboratory world when spectroscopic and chemical data are collected for an unknown structure. A simple example will illustrate this point.

Suppose that the ultraviolet absorption spectrum of an unknown molecule points to the presence of an α,β-unsaturated ketone functionality and the proton NMR spectrum points to a vinyl methyl group. Without additional information, and as long as the molecular formula allows, the double bond in the two substructures may or may not be the same. Therefore, the chemist or computer program must consider as alternative substructures either **141a** and **141b**—both of which assume the double bond to be overlapping between the two substructures—or the pair **142** and **143**, which assumes the double bonds to be distinct. As exemplified with concrete cases,[208,220] GENOA produces an exhaustive

$$-\overset{O}{\underset{}{\overset{\|}{C}}}-\underset{\underset{CH_3}{|}}{C}=C- \qquad -\overset{O}{\underset{}{\overset{\|}{C}}}-\overset{|}{C}=\overset{|}{C}-CH_3$$

       141a   **141b**

$$\underset{142}{-\overset{\overset{O}{\|}}{C}-\overset{|}{C}=\overset{|}{C}-} \qquad \underset{143}{-\overset{|}{C}=\overset{|}{C}-CH_3}$$

and irredundant set of structural isomers based on overlapping and alternative substructure inputs, which offer the chemist absolute assurance that all possible structures consistent with the available data have been considered. The problem of constitutional isomer generation, therefore, had been solved.

Applications to Stereochemical Problems in Structure Generation

With the arrival of James Nourse, we were in a position to attack the last deficiency: securing knowledge about the stereochemistry of the structures generated by GENOA. Nourse had the ideal background: a Ph.D. from Caltech under J. D. Roberts and postdoctoral training at Princeton with Kurt Mislow in stereochemistry. He also had a superb mathematical mind, which, among other achievements, led him to produce, while at Stanford, a national best-selling paperback on rapid solutions of Rubik's Cube. (During the height of the Rubik's Cube craze, only the Bible seemed to enjoy higher annual sales; Nourse was the only postdoctoral fellow in my memory who consulted me on tax shelter questions.) After describing[221] an algorithm that for the first time permitted the enumeration and construction of all possible stereoisomers consistent with a given empirical formula, we were able to make available a program named STEREO,[222] which represented a computer-based approach for constrained generation of configurational stereoisomers that could be combined with the constitutional-isomer generation of GENOA. I shall cite only one example of the power of this approach with the most famous empirical formula in organic chemistry: C_6H_6.

Without constraints, there are 217 possible constitutional isomers and 961 possible stereoisomers—a piece of esoterica that no organic chemist could produce unaided. If various constraints, such as the absence of triple bonds or cumulenes within a six-membered or smaller ring, are introduced, 72 constitutionally feasible isomers remain. Again, no chemist could have produced this number with any degree of confidence. Attempting to draw them would be a fitting diversion for a chemist sentenced to solitary confinement for a few months or years.

The STEREO program revealed[222] that, in addition to the 72 constitutional isomers, there are seven capable of stereoisomerism. Again, no chemist could have made such a quantitative statement with any degree of assurance, but the actual number is small enough that the corresponding structures could be generated manually. During a lecture at a Canadian university, I challenged the entire tenured organic chemistry faculty to produce these seven structures. To my surprise, they accepted this challenge and presented me with the correct answer within 4 hours. For readers of this memoir who would like to try this on their own, the answer is provided in our original publication.[222]

Applications to NMR Spectral Interpretation

On the whole, mass spectrometry is not very sensitive to stereochemical variations, and it was, therefore, possible to use many of the DENDRAL algorithms for mass spectral data manipulation, long before the handling of stereochemical information had been solved. In the case of NMR spectral analysis, stereochemical features play a crucial role. This explains why a considerable hiatus occurred between our initial attempts in 1973[216] to use artificial-intelligence techniques as an aid to NMR spectral interpretations and our much more refined approaches in the early 1980s, with which we ended our work in this field.

We first focused on ^{13}C NMR spectra. As part of the Ph.D. thesis of Christopher Crandell and with major contributions from a British postdoctoral fellow, Neil Gray, we described[223] a program for inferring substructural information from ^{13}C data and for ^{13}C spectrum prediction. A key feature was a successful method[224] for encoding the substructural environment of a resonating nucleus out to a four-bond radius about the atom under question, thereby encompassing effects of δ substituents. We also included a numerical ranking system and illustrated this approach through a detailed retroanalysis of the diterpene 144, whose structure could have been largely solved through the type of computer program incorporated in the GENOA scheme. Our encoding system for ^{13}C substructures could also be used[224] to construct a data base of substructure codes and associated ^{13}C resonances.

144

The most frequent use of such data bases is in the interpretation and prediction of ^{13}C NMR spectra, but these data bases are also useful in the verification of assignments of observed resonances to the atoms of a known structure, when the assignments are based solely on analogies to related, previously assigned structures. Such assignments by analogy carry a fairly high risk of error because the available prototype structures may be inadequate or even inappropriate. We documented this caveat by subjecting the reported ^{13}C NMR spectral assignment of tetrachyrin (**145**)[225] to our computerized checking technique. Even though the reported[225] structure **145** is beyond question, because it is based on X-ray analysis, our computer program[224] questioned the correctness of no less than 11 of 20 spectral assignments in this diterpene.

145

With Huldrych Egli, a Swiss postdoctoral fellow, we extended this work to proton NMR spectra by developing interactive computer programs[226] for the creation and maintenance of a proton NMR data base; the prediction of relevant chemical shifts; and finally, the ordering of structural candidates by rank on the basis of a comparison between observed and predicted spectra. These programs could be linked directly to the GENOA and STEREO programs and were able to take into account configuration at stereocenters and double bonds, as well as diastereotopy. In complicated structural problems, the substructural information deduced from such ^{13}C and ^{1}H NMR spectral data is at times insufficient. By coupling inputs from two-dimensional NMR spectroscopy with the interactive GENOA program, we demonstrated in the last paper[227] of our series that the power of GENOA can be refined even further as a tool for structure elucidation.

Marine Sterols and Phospholipids

The steroid circle of my chemical life really started to close in 1965 when I came out of anesthesia after a gallbladder operation. My surgeon, Roy Cohn, proudly displayed some giant gallstones, but I surprised him by asking whether he would dial my laboratory extension. Still somewhat groggy, I asked Jerry Karliner (now at Ciba–Geigy) to pick up my gallstone, to extract it with ether, and to run the mass spectrum of the crude sterol without further purification. Karliner's Ph.D. thesis problem[228] dealt in part with some aspects of sterol mass spectrometry, and I was curious to determine whether the conventional wisdom about cholesterol being the only sterol constituent in gallstones was correct.

The next day I triumphantly informed my surgeon that, according to mass spectrometry, my gallstone was different. It contained, in addition to cholesterol, traces of two higher homologs. Cohn promptly supplied me with more gallstones than I ever wish to see again, and Karliner then showed that all of them displayed mass spectral peaks of such plant sterol components—presumably originating from the diet. I cited this incident as an example of the utility of mass spectrometry at the Second International Congress on Hormonal Steroids, where I reproduced in the text of my plenary lecture[229] the spectrum of "cholesterol isolated from a gallstone of the author's gallbladder." A footnote after this sentence read: "The experimental contribution of Professor Roy B. Cohn (Department of Surgery, Stanford University School of Medicine) is gratefully acknowledged."

Three years later, a similar situation arose when Paul Scheuer of the University of Hawaii sent me a supposedly pure sterol sample from the marine organism *Palythoa tuberculosa* for mass spectral analysis. The spectrum showed that the specimen consisted primarily of cholesterol and its homolog with a methyl group at C-24, but that there was also present a trace sterol of molecular weight 426, which indicated that the substance was either a triterpene or a hitherto unprecedented C_{30} sterol. I encouraged Scheuer to attempt a laborious gas chromatographic separation of this trace sterol, which eventually was identified as the marine sterol gorgosterol,[230] of then unknown constitution. By mass spectrometric analysis of the derived hydrocarbon, gorgostane, we concluded— from our intimate knowledge of the mass spectral fragmentation of steroid hydrocarbons (cf. Figure 2 and structure 122)—that the three extra carbon atoms had to be located in the side chain. When an NMR spectrum revealed the presence of a cyclopropane ring, it was clear that we were dealing with a sterol of unprecedented composition and biosynthetic origin.

On the basis of these spectral conclusions and additional chemical work, we reported in a joint communication[231] with Scheuer and A. J. Weinheimer (at that time at Stanford on sabbatical leave from his position as professor of chemistry at the University of Oklahoma) that we had narrowed the constitution of gorgosterol to the two alternatives **146** and **147**. Within a few months, my graduate student Nicholas Ling (now a professor at the Salk Institute) succeeded in performing a complete X-ray analysis to demonstrate[232] that gorgosterol had the unique absolute stereostructure **148**. This single structure convinced me that a more extensive survey of marine sterols would be justified.

146   **147**   **148**

Structure Elucidation of Novel Marine Sterols

Not even my inveterate optimism would have led me to the prediction that, over a period of 15 years, we would isolate well over 100 new sterols with structures that then had no counterparts in terrestrial organisms; that we would be involved in difficult underwater biosynthetic experiments; that we would enter the, for us, totally new field of phospholipid chemistry; and that in the middle 1980s I would finally be transformed into a chemical monogamist by dedicating my entire research group to one topic: structure elucidation, biosynthesis, and biological function of such marine lipids.

The chief reasons for our rapid progress were that we benefited from our years of experience in mass spectrometry and NMR analysis of sterols and that this work started after two major advances in separation techniques had been introduced: gas chromatography of high-molecular-weight substances (such as sterols) coupled directly with mass spectrometry (GC–MS), and high-performance liquid chromatography (HPLC). In addition, high-magnetic-field (>360 mHz) NMR spectrometers had become widely available. The combination of all these techniques soon permitted the complete structure elucidation of an unknown sterol with submilligram fractions of material. To cite just one example: my Japanese postdoctorate collaborator, Toshihiro Itoh, was able in just a few months to separate and identify 74 sterols—24 of them with new structures—from a single sample of the sponge *Axinella cannabina*![233]

Gorgosterol (**148**), the first naturally occurring cholesterol analog to have undergone bioalkylation at C-22 and C-23, was isolated from soft corals.[234] However, the real gold mine turned out to be the lowest form of animal life, the sponges. Since 1970, several hundred papers (nearly 100 from our own laboratory) have been published on novel marine sterols, thus constituting what may be the last chemical explosion in the steroid field. For space reasons alone, it is out of the question to do justice to these recent studies within the confines of the present autobiographical account, but fortunately, much of the important American, British, Italian, and Japanese work has been reviewed.[235–241] Therefore, I shall call attention only to some of the most unusual structures, because they are needed to illustrate some of the recent biosynthetic and phospholipid studies, which have so far not been reviewed elsewhere.

Primarily through the work of Minale and collaborators in Naples,[235] it was shown that sponge sterols contain not only the conventional cholesterol nucleus (N) but occasionally also the unique 19-nor (M) and A-nor (O) skeletons, which are generated in the sponge by transformation of dietary cholesterol (N) precursors.[242,243] However, what is really unusual about marine sterols, and especially those found in sponges, is the almost bizarre variety of their side chains.

In terms of new and exciting chemical developments, the field of plant and animal sterols had become essentially dormant by 1960[9,12] except for the elucidation of some of the terminal stages in sterol side-chain biosynthesis.[244] Chemically, the important animal and plant sterols are not very diverse, and they can be categorized (except for double bonds and occasional additional oxygen functions) by the general structure 149. By contrast, three type structures—150, 151, and 152—are needed to describe generically the amazing range of side chains that can be encountered among marine sterols.

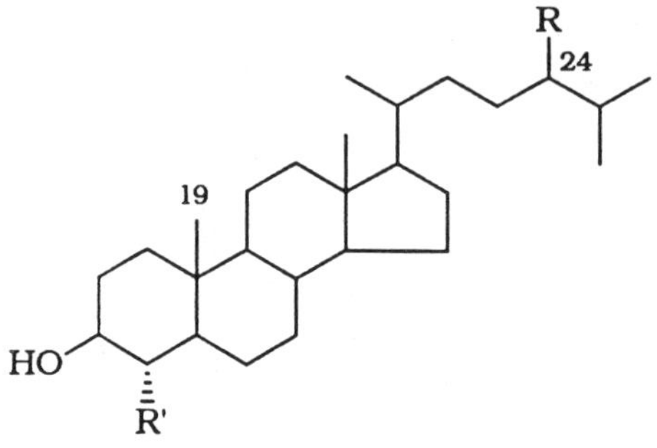

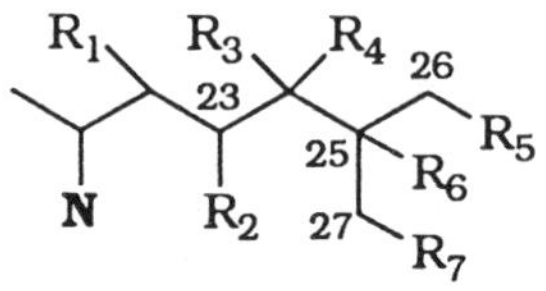

149 R = H, CH$_3$, or C$_2$H$_5$
R' = H or CH$_3$

150

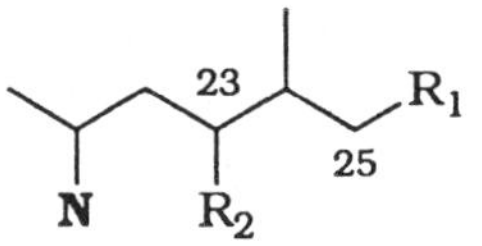

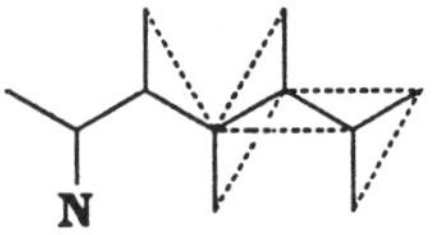

151

152

Whereas the important plant sterols bear only one- or two-carbon substituents at position 24 of the side chain (**149**), marine sterols have been found with alkyl substituents (**150**) at positions 22, 23, 24, 25, 26, and 27 of the cholesterol side chain. In addition, there are truncated variations, of which 24-norcholesterol (**151**, R$_1$ = R$_2$ = H)—as yet of totally unknown biosynthetic origin—is the most unusual example.[245] This unique side chain, frequently with an additional double bond between C-22 and C-23, is so widely distributed in the marine environment as to serve as a geochemical marker. Finally, as foreshadowed by our earlier structure elucidation[232] of gorgosterol (**148**), a plethora of cyclopropane-containing side chains (generic structure **152**) has been isolated, which is also unique to the marine environment and can thus serve as another geochemical index.

In order to illustrate the unprecedented range of bioalkylation of the cholesterol side chain in sponges, Chart I includes eight examples of the many sterols that have been isolated in our laboratory during the past decade. In virtually every case, the structure and relevant stereochemistry was confirmed by synthesis.

Because sterol side chains with a cyclopropyl ring are even more unusual, including even representatives with cyclopropene rings, all such naturally occurring examples are collected in Chart II. Aside from

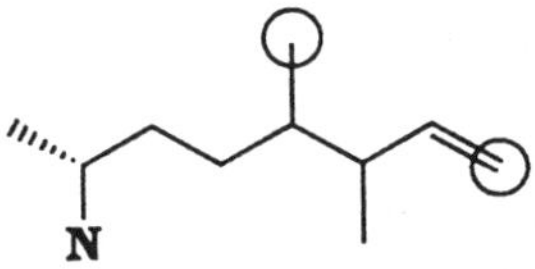

26(29)-dehydroaplysterol 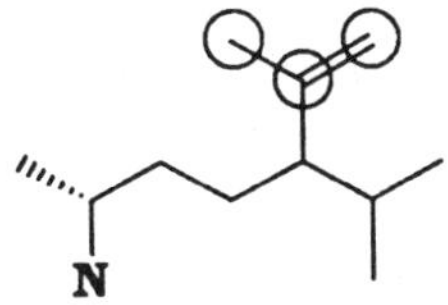   24-isopropenylcholesterol

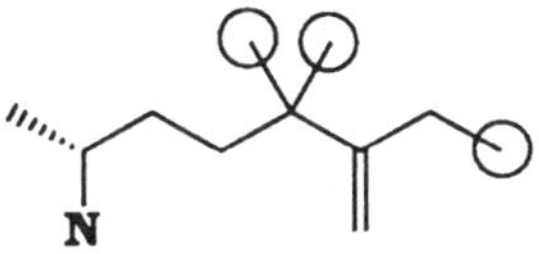

durissimasterol 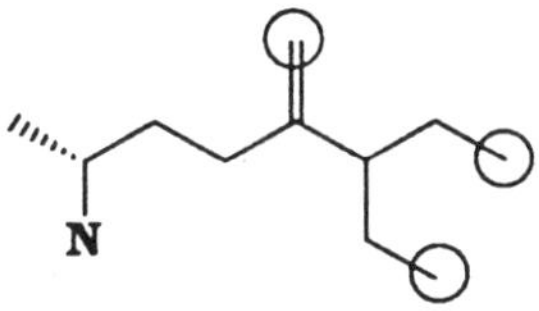   xestosterol

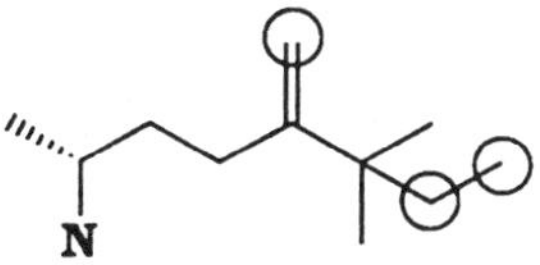

mutasterol 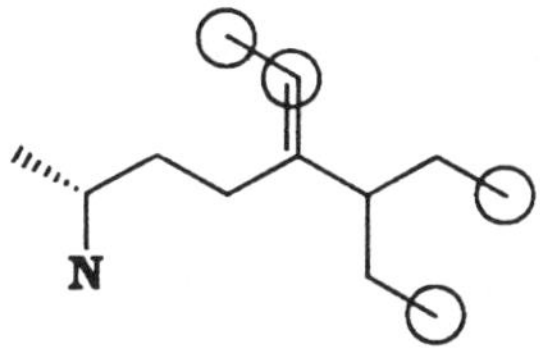   28-methylxestosterol

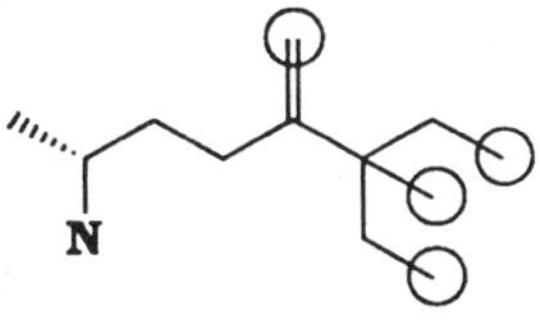

25-methylxestosterol 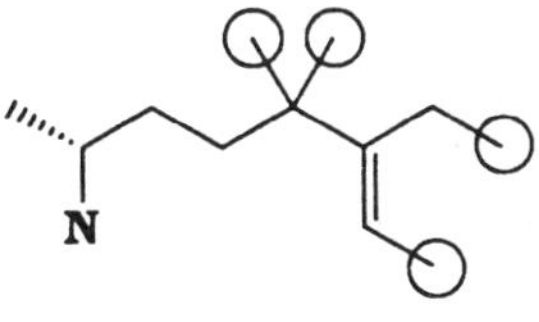   xestospongesterol

Chart I

23-demethylgorgosterol,[246] calysterol,[247] and petrosterol,[248] the other 13 structures reproduced in Chart II were established in our laboratory.[238–240,249–252] With the exception of the calysterols—where synthetic attempts[253] failed, presumably because of the documented lability[254] of the cyclopropene system—all sterol structures collected in Chart II have been confirmed by synthesis.

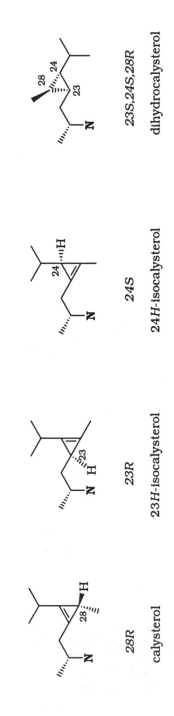

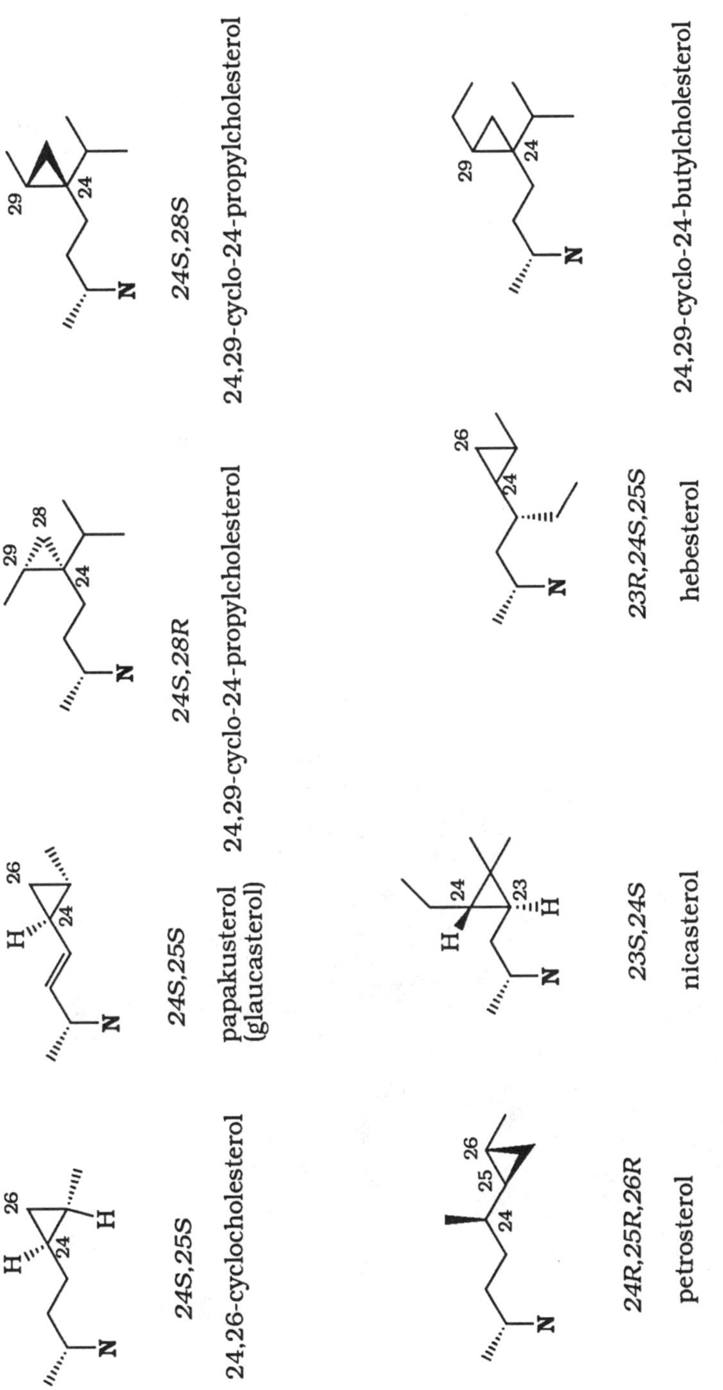

DJERASSI *Steroids Made It Possible*

24S,25S
24,26-cyclocholesterol

24S,25S
papakusterol
(glaucasterol)

24S,28R
24,29-cyclo-24-propylcholesterol

24S,28S
24,29-cyclo-24-propylcholesterol

24R,25R,26R
petrosterol

23S,24S
nicasterol

23R,24S,25S
hebesterol

24,29-cyclo-24-butylcholesterol

Chart II

Biosynthesis of Sponge Sterols

Since 1953[4] we have probably speculated in several hundred papers about the biogenesis of the various alkaloids, terpenoids, and other naturally occurring molecules whose constitutions had been established in our laboratory. Therefore, I have no real explanation of why it took 30 years before we made any serious effort in experimental biosynthesis. Having waited that long, it probably served us right to find that our sirens, the sponges—like many other sessile marine organisms that are filter feeders—are fickle organisms. In addition to practical problems, such as slow metabolic rates and having to perform many experiments far away from any laboratory resources on rafts or under water by scuba diving, there is even the question of whether the sponge or a symbiont is responsible for the actual biosynthetic steps. In soft corals, these problems have not yet been surmounted, and no successful biosynthetic sterol experiment has so far been published.[255] Counter-

My son, Dale, and I are about to go diving in the cold Pacific Ocean at Stanford's Hopkins Marine Station in Pacific Grove, CA.

balancing these complications is that even our occasional failures happened in (figuratively speaking) breathtaking environments near the Great Barrier Reef in Australia and off the coasts of Papua New Guinea, California, or southern Italy. My own initiation into scuba diving occurred on an idyllic island on the north coast of Papua New Guinea, where one of my early Belgian postdoctorate associates, Ben Tursch, had established a small marine biological station. We dove in lukewarm water without wet suits, in marked contrast to my next dive in the kelp forest of the bitterly cold water near Stanford's Hopkins Marine Station.

Once the technical aspects of precursor incorporation were solved by the marine biologist of our group, Janice Thompson (now with the Affymax Research Institute), we undertook the frequently laborious but at times also challenging task of synthesizing the many radioactive precursors that needed to be examined. The synthetic problems were associated with the requirements of high stereochemical purity[256,257] and the fact that the multistep syntheses had to be designed in such a fashion that the radioactive label was introduced as late as possible in the sequence. In addition, for reasons of cost, only very few isotopic reagents were available: ^{14}C-labeled methyl iodide and cyanide, and ^{3}H-containing sodium borohydride or water.

A major role in the early synthetic and incorporation studies was played by a Bulgarian postdoctorate fellow, Ivan Stoilov, who also happened to be an outstanding diver and who spent 3 years in my group as part of a cooperative research program with the Bulgarian Academy of Sciences. Experiments in distant locations (e.g., outer reefs) were performed[256] by collecting the sponge 1 month prior to use and transplanting it onto plastic plaques attached to an underwater grid near the natural habitat. The precursors were incorporated via 12–24-h aquarium incubations[257] (sometimes performed in gallon jars on a raft with aeration from scuba tanks) and then returned to the sea for about 1 month before being collected. On more than one occasion, the sample was lost during stormy weather!

Single and double biomethylation at C-24 of the cholesterol side chain (see 149) by means of S-adenosylmethionine (SAM) has been demonstrated in plants,[244] but the unusual features of the alkylated side chains of sponge sterols had not been subjected to experimental biosynthetic scrutiny. The circles in the structures collected in Chart I denote the extra carbons that are introduced by SAM biomethylation. Until recently,[259] examples of triple (e.g., mutasterol[260]) or quadruple (e.g., 28-methylxestosterol[261]) biomethylation have been unique to marine sterols. Even double biomethylation associated with side-chain extension,[235] as in 26(29)-dehydroaplysterol,[262] had not been encountered before. Both expected and unexpected processes occur in sponge sterol biosynthesis.

We developed our incorporation technique with the Californian sponge *Aplysina fistularis* by showing[257] that its principal sterol, 25-dehydroaplysterol (153a), undergoes extension of the cholesterol side-chain terminus in the expected fashion[244] by SAM bioalkylation of epicodisterol (154a), which possesses the same C-24 stereochemistry as 25-dehydroaplysterol (153a). The radiolabeled epimer, codisterol (154b), was not utilized by the sponge. By contrast, extension of both ends of the cholesterol side chain, as in xestosterol

In contrast to the cold California water, diving in the Indian Ocean was a pleasure and barely required a wet suit. Here I am clowning with the remnants of a lobster aboard the Soviet research ship Akademik Oparin *in the Maldives (1989).*

(155), lacked[258] stereoselectivity because both isomeric codisterols 154a and 154b were now equally well utilized, as were the intermediate 25-dehydroaplysterol isomers 153a and 153b.

Particularly interesting is the biosynthesis[256] of strongylosterol (159), which displayed an unexpected stereochemical feature. The absolute configuration of strongylosterol had been documented[263] in terms of stereostructure 159, and so the a priori most obvious biosynthetic route[244] should be via SAM alkylation of epiclerosterol (158) by analogy to the earlier[257] demonstrated origin of its lower homolog (154a→153a). However, when labeled epiclerosterol (158) was fed (off the north coast of Papua New Guinea) to the sponge *Strongylophora durissima*, no incorporation was encountered. Instead, bioalkylation occurred first by ring extension at C-26 through an amazingly stereospecific utilization of codisterol (154b), but not epicodisterol (154a), followed by further non-stereospecific SAM methylation at C-28 of either epimer 156a or 156b of 24(28)-dehydroaplysterol. This investigation[256] also represents a good example of the major synthetic effort that had to be expended in order to develop feasible stereoselective approaches to the pure labeled precursors 154a, 154b, 156a, 156b, and 158, through which these unexpected stereochemical features of the biosynthesis of 159 were unraveled.

In addition to our biosynthetic studies of side-chain alkylation, we have also paid attention to cyclopropane ring formation. We first focused on petrosterol (161), the principal sterol[248] of the Mediterranean sponge *Petrosia ficiformis*. On the basis of everything that we have learned[244,256–258] about sterol side-chain biosynthesis, this product of a double SAM bioalkylation of the cholesterol side chain should be generated[238] by cyclopropane ring formation of a carbonium ion intermediate (160) derived in one step from SAM attack on epicodisterol (154a), which possesses the same stereochemistry at C-24 as petrosterol (161). To our surprise, labeled epicodisterol (154a) was not utilized by the sponge, whereas 28-^{14}C-24-methylenecholesterol (162)—a totally unexpected precursor—was incorporated with high efficiency.[264] Through extensive degradation of labeled petrosterol, the operation of a complex rearrangement process (163→161a) was uncovered, in which the original labeled C-28 of the precursor 162 ended up at C-24 of the final product 161a. Therefore, we ascribed[265] a crucial role to a protonated intermediate of dihydrocalysterol (163) and subsequently[266] showed dihydrocalysterol (in unprotonated form) to be also the biosynthetic precursor in cyclopropene formation to 24H-isocalysterol (164)[267] and thence, by isomerization, to the other two naturally occurring cyclopropenes, calysterol (165)[247] and 23H-isocalysterol (166).[268]

It is unlikely that any chemist, however experienced and sophisticated in steroid biogenesis, would have predicted the biosynthetic origin of petrosterol (163→161a) or, for that matter, some of the stereochemical features encountered with strongylosterol (159). Just as the diversity of the marine sterols has presented an unexpectedly late challenge to structural work in the sterol field, so some of their biosynthetic idiosyncracies—including the possibility that certain unicellular organisms may be functioning as symbiotic sterol factories—indicate that a great deal can still be learned in that area as well.

Phospholipid Studies

It is only appropriate that I end this chronological account of my research interests with an area, phospholipids, that was totally foreign

to me until 1980. My group's entry into this field was again prompted by steroids, this time by our concern about the possible biological role of the unique marine sterol structures that have been discovered since the early 1970s.

In our first detailed speculation[269] about the biological role of those marine sterols that are present in substantial amounts and virtually unaccompanied by conventional sterols (such as cholesterol), we discussed their possible role as functional components of plasma membranes. Aside from its metabolic role, cholesterol is known[270,271] to increase membrane fluidity and stability and to modify membrane permeability by interaction with membrane phospholipid molecules. Sterols with synthetically extended side chains do not function well[272] in mammalian-type model membrane systems composed of phospholipids typically found in nature associated with cholesterol. This suggests that the natural marine sterols with extended side chains may occur in membrane systems composed of phospholipids that differ from the ones found in mammalian membranes. This hypothesis led us to investigate the phospholipids of sponges with major emphasis on the fatty acid components, because the unusual feature of marine sterols—the highly substituted side chain (e.g., 150)—would be located in the lipophilic portion of the phospholipid bilayer among the fatty acid chains.

Our very first entry into the field, as reported[273] in a portion of the Ph.D. thesis of Robert Walkup (now on the faculty of Texas Tech University), indicated that a relationship between unusual sterol and fatty acid side chains might indeed exist. We selected the sponge *Aplysina fistularis* because its principal sterols, 25-dehydroaplysterol (153a) and its 25,26-dihydro analog, contain the unusual extended cholesterol side chain observed only in marine organisms. The major acyl components of the phospholipids consisted of 85% C_{14}–C_{20} acids (several of them branched, as is common among bacteria) and 15% C_{27}–C_{30} demospongic acid types.[274,275] Table II contains descriptive and quantitative information on the five new fatty acids isolated and characterized by Walkup[273] from this demospongic acid fraction.

Three features merit comment: All of the acids possess the unique 5Z,9Z-diene system (*see* circles in structures in Table II) first detected by Litchfield and collaborators[274] in sponges. All of the acids contain another unusual structural feature: unsaturation or branching five to seven carbons away from the chain terminus. Finally, the bulk of these acids is found among the phosphatidylserine (PS) and phosphatidylethanolamine (PE) classes, rather than the conventional phosphatidylcholine (PC). A rough separation of the sponge into bacteria-enriched and sponge cell fractions showed that the demospongic acids (Table II) are indeed sponge cell constituents.

The branched nature of the principal demospongic acid in *Aplysina*—22-methyl-5,9-octacosadienonic acid (abbreviated as 22Me-$\Delta^{5,9}$-28:2)—pointed toward possible chain elongation from a bacterial precursor. Furthermore, in view of the known[271] disordering effect of the side-chain alkylation of yeast sterols on phospholipid bilayers, both the methyl substituent at 22-C in the demospongic acid (Table II) and the extra methyl groups in the side chain of the aplysterol sterols (*cf.* 153a) may exist to promote increased fluidity in the sponge membrane. Indeed, inspection of molecular models suggests that if the unique 5Z,9Z-diene system were stacked with the nodal planes parallel to allow maximum $\pi-\pi$ overlap, the methyl branch at C-22 of the acid and the terminal groups of the sterol side chain of 153a would be very close. Disordering would then increase.

This hypothesis prompted one of my long-time collaborators, Eser Ayanoglu, as well as several other members of my group, to examine[276-279] the phospholipids of a variety of sponges that were known to contain unusual sterols. Just as with our earlier sterol sponge work, we encountered numerous unusual fatty acid structures that were unknown among terrestrial organisms. A few of these are listed in Table III, with the unusual structural features again emphasized by circles.

By now we were hooked, and our phospholipid work branched out into various directions with implications quite beyond the narrow confines of marine natural products research.

Demospongic Acid Biosynthesis

Because we had solved (vide infra) with Janice Thompson and Ivan Stoilov most of the technical problems associated with the administration of labeled sterol precursors to sponges, we decided to utilize these techniques for an examination of the biosynthesis of these unusual fatty acids. The only relevant work is that of Litchfield and Morales,[275] who proposed, on the basis of work with labeled acetate, that the straight-chain demospongic acids arise from a chain elongation starting with dietary or symbiont precursors such as palmitic (16:0) and palmitoleic (Δ^9-16:1) acids. Their proposed pathway (acetate → 16:0 → 26:0 → Δ^9-26:1 → $\Delta^{5,9}$-26:2), though plausible, required firmer documentation in terms of specific precursors. Acetate is the unit used in all fatty acid chain elongations, and they examined no other labeled precursor in the sponge *Microciona prolifera*. Therefore, we undertook the task of synthesizing a variety of labeled fatty acid precursors and examining in detail the biosynthesis of straight-chain as well as branched demo-

Table II. Novel Fatty Acids from the Phospholipids of *Aplysina fistularis*

Acid	Abbreviation	Distribution in Phospholipid Classes				
		PI	PS	PC	PG	PE
20-Methyl-5,9-hexacosadienoic	20Me-Δ^{5,9}-26:2	—	2.2	—	—	2.4
5,9,21-Octacosatrienoic	Δ^{5,9,21}-28:3	1.0	8.3	—	—	3.6

Structure	Name	Shorthand					
5,9,23-Octacosatrienoic	$\Delta^{5,9,23}$-28:3	—	3.1	—	—	3.0
22-Methyl-5,9-octacosadienoic	22Me-$\Delta^{5,9}$-28:2	1.6	20.0	0.6	1.1	17.8
5,9,23-Nonacosatrienoic	$\Delta^{5,9,23}$-29:3	—	1.4	—	—	1.2

Table III. Unusual Fatty Acids from Some Sponge Phospholipids

Acid	Abbreviation	Source
24-Methyl-5,9-hexacosadienoic	24Me-$\Delta^{5,9}$-26:2	*Petrosia ficiformis*[276]
2R-Methoxy-21-octacosenoic	2MeO-Δ^{21}-28:1	*Higginsia tethyoides*[277]

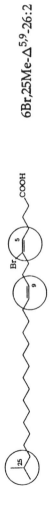

6Br,25Me-$\Delta^{5,9}$-26:2   *Petrosia hebes*[278]

6-Bromo-25-methyl-5,9-hexacosadienoic

19Me,24Me-Δ^5-25:1   *Strongylophora durissima*[279]

19,24-Dimethyl-5-pentacosadienoic

iso-15:0 ⟶ $\Delta^{5,9}$-iso-27:2

antelso-15:0 ⟶ $\Delta^{5,9}$-antelso-27:2

n-16:0 ⟶ $\Delta^{5,9}$-26:2

spongic acids. As shown below, we chose the right substrates to uncover a number of interesting aspects of fatty acid biosynthesis.

We first addressed the question of branching, because such branched long-chain fatty acids had not been described before in the literature. The anteiso- or iso-terminal chain branching exemplified by the first and third entries of Table III are typical of bacterial acids,[280] except for the fact that the latter are of shorter chain length (C_{13}–C_{23}). Our first choice was the Australian sponge *Jaspis stellifera*, whose predominant sterols (128 and 156) are of the unusual marine type (150). Its three most abundant long-chain fatty acids are the most widely distributed sponge acid,[274] 5Z,9Z-hexacosadienoic acid ($\Delta^{5,9}$-26:2) and the two branched homologs, 25-methyl-5,9-hexacosadienoic acid ($\Delta^{5,9}$-iso-27:2) and 24-methyl-$\Delta^{5,9}$-hexacosadienoic acid (Δ5,9-anteiso-27:2). By synthesizing the appropriate ^{14}C-labeled precursors, Nestor Carballeira (now on the faculty of the University of Puerto Rico) could show[281] that these demospongic acids were produced by chain elongation of the short-chain 16:0, iso-15:0, and anteiso-15:0 precursor acids, which are typical of bacteria.

A similar situation was encountered[282] in the principal demospongic acid of *A. fistularis* (Table II), the 22-methyl-$\Delta^{5,9}$-28:2 acid, where we also attacked the question of the chirality of the branching site. As part of his Ph.D. thesis, Arthur Shu[282] accomplished the total synthesis of both R and S isomers, starting with terpenoid precursors of known absolute configuration, and showed that the sponge acid was the 22R-(−)-enantiomer. This work was then extended by a French postdoctoral fellow, Daniel Raederstorff, who made the surprising observation[282] that 10R- as well as 10S-methylhexadecanoic acid (10-Me-16:0)—an acid that we had isolated from bacteria present in that sponge and then synthesized in optically active form—are homologated in vivo. It appears that the asymmetric center is sufficiently distant from the carboxyl group so that the sponge's elongase enzyme system does not discriminate efficiently between the two antipodes.

Having demonstrated that the sponge utilizes bacterial and possibly other short-chain precursor acids for chain elongation to the unusual range ($>C_{26}$) of the demospongic acids, Soonkap Hahn[283a] in my laboratory then fed a variety of specially synthesized precursors to the sponge in order to clarify the manner in which the double bonds are introduced. His results are summarized in the following scheme.

$$16:0 \rightarrow 26:0 \rightarrow \Delta^9\text{-}26:1 + \Delta^5\text{-}26:1$$

$$\Delta^9\text{-}16:1 \rightarrow \Delta^{19}\text{-}26:1 \rightarrow \Delta^{5,9,19}\text{-}26:3 \leftarrow \Delta^{5,9}\text{-}26:2$$

Particularly striking is his observation that both the Δ^5-26:1 and Δ^9-26:1 acids are efficiently transformed to the same $\Delta^{5,9}$-26:2 product, but that the sponge's Δ^9-desaturase is incapable of acting at the short-chain 16:0 level. This is contrary to the standard assumption[275,284] that Δ^9-desaturation always precedes the introduction of further double bonds in the biosynthesis of polyunsaturated fatty acids. In addition, Hahn et al.[283b] found that while freshwater sponges also contain the same demospongic acids as marine sponges, some of the biosynthetic desaturation sequences differ significantly. Work is currently underway in our laboratory to determine whether these desaturations in marine and freshwater sponges occur at the conventional[284] coenzyme-A stage or whether desaturation can also proceed in the sponge phospholipid.

Phospholipid Characterization, Localization, and Liposome Formation

For a rational attack on the possible functional role of marine sterols in plasma membranes through their interaction with phospholipids, it was essential to deal with pure molecular species. Consequently, a substantial effort was expended on effective procedures for separating and purifying individual phospholipids with unusually long acyl chains[285] and on mass spectrometric means (e.g., desorption chemical ionization and fast-atom bombardment)[286] for their characterization. This investment into methodological advances proved to be more than justified when we discovered[285] that most of the demospongic-acid-containing phospholipids bear identical acyl substituents in the glycerol moiety, in contrast to virtually all animal phospholipids,[287] in which the two acyl groups are different.

The next step was to undertake model membrane studies with these unusual phospholipids and to examine the influence of different sterol structures upon their biophysical properties. For this purpose, we did not wish to depend solely on naturally occurring phospholipids, especially because we also wished to examine the significance of the unusual 5Z,9Z-diene system in the demospongic acids. Therefore, stereospecific syntheses[288,289] were developed of the four possible isomers (5Z,9Z; 5E,9E; 5Z,9E; and 5E,9Z) of these acids and their conversion into the appropriate phospholipids with various head groups.[289,290]

Our initial model membrane studies,[291] conducted in collaboration with Nejat Düzgünes of the University of California Cancer Research Institute, by means of differential scanning calorimetry and light-scattering measurements, gave totally unexpected results. Liposomes composed of either 1,2-di-(5Z,9Z)-5,9-hexacosadienoyl-*sn*-glycero-3-phosphocholine or ethanolamine underwent an endothermic phase transition at 42 °C, which was not affected by either cholesterol or several marine sterols with unusual side chains. In fact, all of these sterols were excluded from the phospholipid bilayers! This result, of course, was contrary to our working hypothesis[269,273] and might be explained most easily by the assumption that the marine sterols are not present in plasma membranes. However, detailed work in our laboratory[292] involving cell separations, plasma membrane fractionation, and electron microscopy did verify that both the unusual sterols and the demospongic-acid-containing phospholipids are sponge plasma membrane constituents. Finally, the recent availability[289] of a series of synthetic phospholipids, with acyl groups in the range C_{18}–C_{24} and containing the unique 5Z,9Z-diene system, has shown that sterol incorporation by such liposomes is strongly affected by chain length. It is clear that a great deal of synthetic and biophysical work, coupled with

I often use this picture (1988) from one of the skiing trips with my students in lectures to point out that while they are active, I relax and just report their data.

careful cell separations, still remains to be performed before a definite judgment can be made about possible interaction between these unusual sterols and demospongic acids in sponge membrane phospholipids. In addition, certain membrane proteins may be affected in some special manner by these unique marine sterols, similar to the known[271] cholesterol–protein interactions in some animal tissues. Work has been initiated in our laboratory to determine whether sponge membrane proteins[293] have unusual features that would justify such a hypothesis.

It is conceivable that such research with the lowest form of animal life may contribute to our knowledge of cell membrane function in higher organisms. As demonstrated in this chapter, we would never have become involved in such an endeavor had it not been for our interest in the mass spectrometric fragmentation behavior of steroids, which in turn led us, via gorgosterol (164), to a broad survey of marine sterol structure and biosynthesis. Thus, even with phospholipids, I was fortunate to have worn my steroid glasses.

A different set of glasses was used by Ben Tursch, a Belgian former postdoctorate fellow of mine, when he wrote an "article" (*see* pp 140–143) on the occasion of my laboratory's "milli-pub" celebration.

Epilogue

Most active scientists are so enmeshed in their current work that they pay little attention to what they, or others, did in the past except for material that may be relevant to their current research. Now, for the first time in decades, I find myself reviewing and judging the entire body of my scientific work. Initially this activity was exhilarating, but in the process I have gradually become preoccupied with a potentially dangerous question: Was it worth it?

The quick and superficial reply would have to be "Yes, of course." But longer reflection during the weeks it took me to write this volume leads to a more complex answer, which I shall now present in terms of the two legacies that any scientist leaves: human progeny and scientific facts.

Students

This chemist's progeny are the several hundred graduate students and postdoctorate fellows from over 50 countries who have worked with me over the years. The work they have done in academia, industry, or government since leaving my laboratory certainly has made my own academic career worthwhile. One of the most enjoyable aspects has been receiving in my laboratory chemical "grandchildren" who have been formed by previous students of mine and who have now come to Stanford for further education.

I have not mentioned undergraduate students because I have taught virtually no undergraduate chemistry. My only direct influence has probably been on the seniors and juniors (and occasional sophomores like Harvey Lodish,[294] now a professor of biology at MIT) who performed undergraduate research in my laboratory. Yet all of my teaching for the past dozen years has been at the undergraduate level, though not in chemistry. My ever-increasing concern with our society's scientific illiteracy and concurrent propensity for seeking oversimplified, black-and-white answers to inherently complex and gray questions has made me become deeply involved in one of the most popular undergraduate majors at Stanford, the Human Biology program. As a member of its executive committee and as the only chemist on its

Aboard Ben Tursch's boat in Papua New Guinea. Ben (right), Diane Middlebrook (now my wife), and I (in background) were on the way to the island of Boisa.

Tetraodon Letters No 37, pp 1839-1843, 1982.
Pergamon Press Ltd. Printed in Great Britain

MINOR AND TRACE STEROLS IN MARINE INVERTEBRATES. Part 2.79 x 10^2.
18,19-BISNOR-(4,5),(9,10),(8,14),(13,17)-TETRASECOPREGNANE,A NOVEL
STEROID FROM THE SPONGE Aristospongia caveatemptor (Meereslustig,1892)[1]

J.E.K.CRABMAN, D. DE ZOLA, A.N. KISMI, N. SCHUBERT

Collectif de Bio-Ecologie, Faculté des Sciences
Université Libre de Bruxelles - 1050 Bruxelles - Belgium.

and

Carl DJERASSI

Department of Chemistry, Stanford University, Stanford, California 94305.

(Received in UK 6 December 1979, accepted for publication 10 March 1982)

During our continuing search for novel marine sterols we have repeatedly documented[2] the presence of unusual sterol derivatives in sponges. The climax of this ex(t,p)ensive program was recently reached with the rediscovery of the title animal in New Guinea waters. The sponge *Aristospongia caveatemptor* (Porifera, Anopsispongidae), although apparently not uncommon around Laing Island, is exceedingly difficult to collect and indeed its very existence has been doubted until recently[3]. As stated in the original description[4], the colourless sponge very exactly matches the refractive index of the surrounding water and is thus unobservable for predators (and collectors) acting on simple visual cues. This remarkable adaptation is very probably due to highly specialized cell membranes microstructures, suggesting the likely presence of unusual sterol material.

An adequate methodology directly derived from the principles of optical rotatory dispersion was developed[5] in order to collect the elusive sponge. Continuous eyelid blinking induced in divers by liberal addition of stoichiometric mixtures of V.S.O.P.[6]/ S.P.[7] resulted in subliminal differential measurements between left and right eye fields at unusual wavelengths, thus permitting visualization of the sponge[8]. Since all vessels containing extracts or purified compounds of the sponge naturally appear empty *(vide supra)* the same shift reagent had to be continuously used by all members of our research teams until completion of this work.

The collection of the sponge documented a well-precedented attractive bonus, namely travel[9]. *Aristospongia caveatemptor* was gathered in 2.05 feet of water at Laing Island in the second week of September 1979 during a stiff breeze, as much as we remember. The sponge proved to be another sterol gold mine[10].

1839

The "article" that Ben Tursch wrote for my laboratory's "milli-pub" celebration.

1840

A methylene chloride extract of the sun-dried material, followed by repeated silicagel column chromatography afforded an auspicious fraction that was shown by tlc to consist of a mixture of essentially pure compounds. This was further fractionated by HPLC and reverse-phase chromatography, yielding a small amount of pure, colourless and invisible compound I (oil ?; $(\alpha)_0^{25}$ 0.002 $\pm$ 15°; UV (in hexane) : end absorption only ; IR : 2850, 2735, 1470, 1375 and 820 cm^{-1}; NMR : complex signal (35 H) centered at 1.26, ill-resolved absorption (6H) at 0.91 ppm -several strong signals totaling 107 H were obviously due to small impurities and thus discarded). The structure was established in a remarkably short time. The most informative information was provided by the MS, apparently consisting exclusively of metastable ions. These results were beautifully supplemented by the amazing S.M.I.P - M.S. technique [11], establishing the probable existence of a nearly-molecular quasi-ion M+1 at m/e 269 (HRSMIPMS : 269.3195 ; calc. $C_{19}H_{41}$: 269.3207). The presence of 19 carbon atoms immediately suggested the close relationship of compound 1 with 18,19-bisnorpregnane. The corpus of spectral data indicated the presence of two terminal methyl groups and 17 methylene groups. Intensive use of artificial intelligence (CONGEN-GENOA) reduced the number of possibilities to one : 18,19-bisnor-(4,5),(9,10),(8,14),(13,17)-tetrasecopregnane (I). The remarkable chemical reactivity of compound 1 (acetylation, oxidation, hydrogenation and Bischler-Napieralski — modified according to Tchitchibabine — were all negative) is in full agreement with the proposed structure.

The structure, stereochemistry and absolute configuration of compound 1 were unambiguously established by direct chemical correlation with pregnane. The epoch-marking simultaneous discovery in Stanford and Brussels that under certain experimental conditions a wide variety of complex marine sterols could be transformed into silicone grease and butyl phtalate in yields of over 100% had previously led to the isolation and description of the remarkable enzyme steroldecyclase [12]. This enzyme provided the basis for an elegant correlation between our compound and the classical steroid series.

Continued on next page.

1841

3.756 kg of pregnane (2) [13], submitted to very mild Kuhn-Roth conditions, yielded 0.17 mg (thus justifying fully the title of this Series) of the key degradation compound 18,19-bisnorpregnane (3), accompanied by some impurities. The crude reaction mixture was immediately incubated at 29.04 ° C, pH 5.695 with steroldecyclase [12] affording in good yield, after purification, compound 1, undistinguishable by all its chemical properties from the invisible natural product.

Of considerably greater scientific relevance were the results of the treatment of compound 3 with steroldecyclase in the presence of tritiated seawater. Total absence of isotope incorporation completely confirmed our personal views on the reaction mechanism[13]

With our previous studies on minor and trace marine invertebrates [14] steroid chemistry had reached an absolute summit : it has now taken a gigantic step forwards. Indeed, 18,19-bisnor-(4,5),(9,10),(8,14),(13,17)-tetrasecopregnane is the first steroid entirely devoid of rings and consists exclusively of an unprecedented nineteen carbons side-chain. Biogenetically, compound 1 is the long-awaited missing link between methane and xestospongesterol. The complete lack of chirality in the molecule suggests a very great biochemical plasticity and makes it a choice candidate for interesting incorporation experiments. It is tempting to speculate that the associated phospholipids of the cell membranes of A.*caveatemptor* will also reflect the dramatic evolutionary changes evidenced in compound 1 and might even be devoid of phosphorus.

Preliminary data indicate bright possibilities for the use of 18,19-bisnor-(4,5),(9,10)-(8,14),(13,17)-tetrasecopregnane in human contraception. A dramatic decrease in the usual number of pregnancies was reported by female research workers of both teams who volunteered to hold at appropriate moments a solid pill preparation [15] of compound 1 tightly between the knees. Further investigations along these lines are actively pursued in our laboratories.

Acknowledgements. This work was generously supported by the National Institute of Wealth (grant JB-007) and the Belgian Institut pour l'Enragement de la Recherche Scientifique dans l'Industrie et l'Agriculture (I.R.S.I.A.). One of us (A.N.K) gratefully aknowledges the receipt of a Fulblast fellowship. N.S. thanks the Institut Royal des Sciences Surnaturelles for the loan of a diving mask (n° ARS/3764-6743/NG/2098658). We are grateful to Dr. Lois Durham who would have recorded the 360 MHz nmr spectra, had she been given any material. We thank Prof.G.Sourison *(ut eructant quirites)* for lack of comments.

Ben Tursch's "article", continued.

1842.

References.

1. This paper was presented at the 1982 convention celebrating C.D's 1000th publication. This is paper $1000 + n$, $(n \to \infty)$.
2. See this Series, parts 1.17×10^2 to 2.78×10^2.
3. Z. von Spielvogel, *Proc.Mick.Mouse Soc.*, 78, 382, (1980).
4. L. Meereslustig, *Dreizig Jahren unter Kannibalen*, Leipzig, 1892, p.1547.
5. J.Pierret, N.Schubert, D.Middlebrook and Carl Djerassi, *Trans.Hansa Bay Philos.Soc.*, Series K, 182 (211), pp. 1-981, (1980).
6. Supplied by Courvoisier, Ltd.
7. Supplied by South Pacific Breweries Ltd., Port Moresby.
8. N.Schubert et al., *J.Chem.Necrol.* 18, 211, (1981).
9. Carl Djerassi, *Pure and Applied Chemistry*, 41, 113-144, (1975) —unexpurgated version.
10. Carl Djerassi, *Pure and Applied Poetry*, in press.
11. Scrambled Metastable Ions Partition Mass Spectrometry. See : J.Smith, K.Smith, O.Jones, W.Jones, A.Dupont, W.Dupont and Carl Djerassi, *Organic Mass Spectrometry*, 12, 432, (1980). For applications, see M.Borboleta, J.Papillon, K.Schmetterling, A.Moth and Carl Djerassi, *J.Infect Physiol.* 71, 123, (1981).
12. J.E.K.Crabman, D. de Zola, A.N.Kismi, N.Schubert, J. Vastifar Hadji, J.Smith, K.Smith, K.Bouter IV Jr., A. Kichi Duoduma, Kristina Katzenellenbogen née Finkelstein, R.Pinto del Pinar Martinez Rosas Pulque Bamba y Mujeres Stop and Carl Djerassi, *Spheroids*, 31, 231, (1979).
13. Carl Djerassi in *Marine Unnatural Products*, P.J.Scheuer, Ed., Vol. 26, in preparation.
14. See This Series, Part 2.77×10^2.
15. Patents pending.

Since 1980, I have also given some courses in Stanford's Feminist Studies Program. My interest in feminist issues surprised my hosts on a lecture trip to Taiwan in 1989. Here I am shown with the leading members of the fledgling Taiwanese feminist organization called Awakening. Left to right: S. C. Huang, C. R. Bo, W. Peng, C. L. Wu, Y. C. Lee, and C. H. Cheng.

faculty, I have taught various policy courses, ranging from "Feminist Perspectives on Birth Control" or "Biosocial Aspects of Birth Control"—courses that have changed the professional career goals of numerous students—to "Pest Control: Technical and Policy Aspects", and given various lectures on health policy.

Public Policy Issues

These teaching activities, talks to lay audiences, and appearances on radio and TV during the past 15 or more years reflect my feeling that scientists must justify to themselves and to the public what their work is all about. As a person who has worked probably longer than most others *simultaneously* in two worlds—the academic and the industrial— and who has always been interested in practical applications of his work, I feel that I am in a particularly good position to speak out in an informative way.

I have also done this in some of my publications. "Birth Control after 1984"[295] has undoubtedly been the most influential public policy paper that I have ever published, because it documented for the first time that our policymakers, and even many scientists, were ignoring the consequences of the extremely long lead times (up to 20 years) required to convert a laboratory discovery into a practical birth control method usable by millions. That article has been reprinted numerous times, and my dismal prognosis[296] about advances in birth control for the balance of the century unfortunately turned out to be well founded. I have continued to lecture and write on this topic and will cite here only four papers[297-300] that I have written on this subject since 1987, as well as the text[301] of my most recent Commonwealth Club address. As an outcome of one of my Human Biology courses, I published with Andrew Israel (then a Stanford Human Biology undergraduate and now a physician in private practice) and Wolfgang Jöchle (then director of veterinary research at Syntex) the first critical study[302] of the "pet population problem" in the United States.

For a period of 15 years, concurrently with my professorship at Stanford, I served as chief executive officer of Zoecon Corporation (since 1983 a wholly owned subsidiary of Sandoz), a company founded in 1968 in the Stanford Industrial Park for research on new methods of insect control based on insect hormones. My interests in new approaches in the pesticide field led me, together with another former Stanford Human Biology undergraduate, Christina Shih, and John Diekman (a former Ph.D. student and then Zoecon's vice president for research and development), to write a major policy paper[303] entitled "Insect Control of the Future: Operational and Policy Aspects." This article probably played some role in the Environmental Protection Agency's (EPA) subsequent revision of its registration policies to permit more rapid examination of certain "biorational" chemicals for which we had offered both a definition and a societal rationale.

Activities in Third World Countries

My first sojourn at Syntex in Mexico around 1950 left an indelible impression on me in an important respect: an awareness of the huge gap between the scientific "haves" and "have nots." In terms of my scientific upbringing, I am totally American. Yet during those first 2 years in Mexico, I started to feel like an outsider who was going to show the *gringos* that first-class research could also be conducted in a country without a great scientific tradition, or indeed in one that had then virtually no Ph.D. chemists.

Concurrent with my Syntex research, I became heavily involved with the budding Instituto de Quimica of the National University of Mexico (UNAM). All of the Institute's research expenses were borne by Syntex, and most of the original staff members as well as its first three directors were trained in one way or another at Syntex. Of all my honorary doctorates, my very first one at age 29 from UNAM has been the most meaningful. It included a hilarious incident—even commented upon in the Mexican press—when Gilbert Stork had a flashbulb explode in his hand as he tried to photograph the ceremony. There had been some bombings in Mexico City around that time, and everybody responded as if this were another terrorist explosion. Instead of giving an appropriate response to the rector, I broke into uncontrolled laughter on seeing my friend's shocked expression.

During my second Mexican sojourn, starting in 1957, we instituted at Syntex an industrial postdoctorate fellowship scheme and continued it when the Syntex research operation moved to the Stanford Industrial Park. In 1972, when I relinquished my position as president of Syntex Research (held concurrently with my Stanford professorship), over 30 postdoctoral fellows were in residence at any one time.

During one of my many trips to Brazil with Ben Gilbert (left), Ernest Wenkert (now at the University of California—San Diego), and the late Venancio Deulofeu (University of Buenos Aires).

My conviction that first-class research can be done in many locations outside the usual East–West axis extended also to Brazil and Africa. My very first Brazilian postdoctorate fellow, Walter Mors, became a good friend after spending a year at Wayne University as a Rockefeller fellow while on leave from his position at the Instituto de Quimica Agricola in Rio de Janeiro. With Rockefeller Foundation support, we started a cooperative program whereby two postdoctorate fellows, first from Wayne and subsequently from Stanford, would always be in residence in his Rio laboratory. With a group of young Brazilian chemists, they conducted a program on Brazilian natural products with major emphasis on alkaloids.

Among the early members of that group were Ben Gilbert, Keith Brown, and Ben Tursch (subsequently a professor at the University of Brussels); Gilbert and Brown remained in Brazil and have developed important research groups of their own. Several of their young Brazilian colleagues eventually came to Stanford for further training. One of the best was Hugo Monteiro (later chairman of the Chemistry Department of the Federal University of Brasilia). During his Ph.D. days at Stanford he consistently referred to me as *El Supremo*. My new title might have gone to my head were it not for the fact that this appellation was invariably accompanied by a charmingly impertinent Brazilian grin, and that I had lost a $100 bet when I challenged him that a certain experiment could not be completed in 7 days. He even had the gall to request the canceled check from me after confessing that he had won the bet because he had subcontracted the work among three other graduate students who split the money with him.

For over a decade, I made one or two annual trips to Rio. The scientific and human productivity of this program convinced me that such collaboration could be formalized on a larger scale. An opportunity to test this belief arose during my chairmanship of the National Academy's Latin America Science Board, when I proposed the creation of the U.S.–Brazil chemistry program.[304] Its purpose was to establish academic centers of excellence at the universities of São Paulo and Rio—similar to what had been accomplished during the past decade in the natural products field with Walter Mors and Ben Gilbert—in inorganic, synthetic organic, polymer, and experimental physical chemistry. Overall guidance would be provided by senior American professors, who would send postdoctorate fellows to the laboratories of their Brazilian counterparts in order to help in the training of graduate students and the conduct of research in those fields.

I chaired this program for the first few years and convinced a group of senior American chemists, none of whom had ever been to Brazil, to commit themselves to this cooperative effort sponsored by the U.S. National Academy of Sciences and the Brazilian Research Council.

During the 1960s and 1970s, I attended many Pugwash Conferences on Science and World Affairs. Here I sit with Georgi Arbatov, director of the American Institute of the USSR Academy and a major Kremlin advisor, at the Pugwash Conference in Sochi, USSR, in 1969. The picture is inscribed, "We do not look at each other, but it really does not mean anything bad. Do you agree? Arbatov."

Not a single person refused. I suspect that Henry Taube, Harry Gray, William Johnson, Charles Overberger, John Baldeschwieler, and George Hammond, to name only a few, have not regretted their annual visits to Brazil. The chairmanship was subsequently taken over by a distinguished Brazilian–American chemist, Aaron Kupperman of Caltech, and the program has produced a large number of Brazilian M.S. and Ph.D. graduates. In 1985, the chairmen of the chemistry departments in both Rio (Bruce Kover) and São Paulo (José Riveros) were products of that program.

In 1967, at the 17th Pugwash Conference on Science and World Affairs in Ronneby, Sweden, I generalized my Mexican and Brazilian experiences in a paper entitled "A High Priority? Research Centers in Developing Nations," which was subsequently published.[305]

> The question is raised whether a basic research center of an internationally recognized standard of excellence can be created in a country where no such centers exist and where the requisite scientific manpower is not yet available. An

affirmative answer is provided by proposing a model which embodies the following three features: (1) an international cadre of postdoctorate research fellows; (2) overall scientific direction by a group of part-time directors from major universities in different "developed" countries; and (3) selection of research areas with a possible ultimate economic pay-off and a maximum multiplication factor—a typical example from Mexico being cited in support of this thesis.

My proposal, which included a recommended annual operating budget, had a remarkable effect. A distinguished African entomologist, Professor Thomas Odhiambo of Nairobi University, wrote to suggest that Kenya be selected as a test site. After a small meeting with some American scientists at the headquarters of the American Academy of Arts and Sciences (then also the home of the American Pugwash Committee), we organized an initial meeting in Nairobi to which representatives from various foreign academies (e.g., Royal Society of London and Royal Swedish Academy of Sciences) and scientific organizations (e.g., Max Planck Gesellschaft) were invited by the American Academy of Arts and Sciences and the U.S. National Academy of Sciences.

Out of all this came ICIPE—the International Centre for Insect Physiology and Ecology—which has probably been one of the most remarkable examples of international cooperation by academies.[306] Now

With Thomas Odhiambo (seated) at one of our many ICIPE meetings in Nairobi.

With Koji Nakanishi (left, in front of vehicle) and Jerrold Meinwald (center foreground) in preparation for our departure from Nairobi to the Tanzanian border.

a flourishing institution with a multi-million-dollar annual budget, it is still headed by Thomas Odhiambo. Among the original long-distance research directors were two organic chemists: Koji Nakanishi from Columbia and Jerrold Meinwald from Cornell. Among my recollections from numerous African trips remain two memorable ones with these friends: Meinwald, a flutist of Boston Symphony caliber, played Mozart for me on a magically beautiful African night in the woods outside Nairobi, and Nakanishi gave one of his performances of legerdemain before a group of Masai while we were waiting for clearance on the Tanzanian border. It was the only time that I saw him drop a prop. To this day he does not believe that I heard one of the Masai mumble, "That's the trouble with the Japanese."

My other major involvements with Third World Countries has been through the National Academy's Board on Science and Technology in International Development (BOSTID), including numerous bilateral workshops in various South American, African, and Asian countries, and my participation in World Health Organization activities. For many years, I was active on committees of the WHO's Special Programme of Research, Development and Research Training in Human Reproduction. The high point of this activity was the organization of an international chemical synthetic effort on novel injectable contraceptives, which was performed in a variety of developing countries[307] and has now culminated in a clinically promising ester of norgestrel (64). During a 5-year period ending in 1987, I served on the scientific and technical advisory committee of the WHO's Special Programme for Research and Training in Tropical Diseases.

Professional Friends

What about my professional friends, my peers from all these years? I will only mention a few—primarily those I have known for decades and whom I still treasure as friends. First and foremost is Gilbert Stork, my classmate from University of Wisconsin days, a friend since the 1940s. I probably learned more about organic chemistry at Wisconsin during our daily joint lunches than from all the courses and professors. In 1950 I brought him to Syntex as its first chemical consultant (to be followed by Arthur Birch and Franz Sondheimer), and he has maintained that relationship with Syntex until now. One of these days, I might write a book of vignettes about Stork. Considering the number of his students and admirers, such a volume might well become a best seller.

Howard Ringold (extreme left) and I with Syntex consultants in the mid-1960s. Seated are Franz Sondheimer, A. J. Birch, and C. J. Sih. Standing are Gilbert Stork and Koji Nakanishi.

At my 60th birthday celebration with my father, Alejandro Zaffaroni, and George Rosenkranz. I am again on crutches because of a hiking accident.

George Rosenkranz (now board chairman emeritus of Syntex) and Alejandro Zaffaroni (now chairman of the board of Alza Corporation) are two intimate friends from my very early Mexican Syntex days. Indirectly, I am the one who introduced them to Palo Alto. Zaffaroni, a Uruguayan biochemist and then executive vice president of Syntex in Mexico City, became the first president of Syntex Research when the company moved to the Stanford Industrial Park as a result of my own move to Stanford. Rosenkranz's three sons—all of them now Stanford alumni—might never have thought of studying in California were it not for the many Palo Alto board meetings of Syntex Corporation.

I have already mentioned Lederberg, Mislow, and Moscowitz and the impact they had on some of my research, but not yet Derek Barton, who belongs in the same category. I was literally the first chemist Barton met when he arrived in the United States in 1949 on his way to Harvard, just a few days before I departed for my first Syntex sojourn in Mexico City. We saw a fair amount of each other in the 1950s and 1960s, especially when our research interests in the steroid, terpenoid, and stereochemical fields intersected. Not so long ago, we celebrated our 35th anniversary in Seoul by reminiscing in the company of some Korean *gisengs* about the similarity in our respective academic careers. Both of us had been essentially mentorless when we secured our first academic appointments—Barton at Birkbeck College and I at Wayne University—in urban institutions that were fairly far down in the academic pecking order.

Derek Barton (extreme right) and I with Finn Sandberg (University of Uppsala) and two Korean colleagues on the 35th anniversary of our first meeting in Seoul. When we were taken out that evening to be entertained by Korean gisengs, Derek and I were the only participants who refused to sing in public, although we did compromise and dance.

My oldest continuing friend from Wayne University days is Andre Dreiding of the University of Zürich, whom I met in Detroit at the time of his first wife's death. Our friendship has been both personal and chemical. For instance, he was one of the first foreign chemists to use our GENOA and other DENDRAL programs, and for years he maintained a direct computer link between his home and Stanford.

Koji Nakanishi was my host in Nagoya in 1957 on my first lecture trip to Japan. He translated my book on optical rotatory dispersion into Japanese, and I arranged for the English publication of his first book on infrared spectroscopy. Nakanishi became one of the early scientific consultants to Zoecon when I was chief executive officer of that company, and our paths still cross in Palo Alto and at international conferences. I feel that at more than one occasion I was the impresario for his magic performances. Perhaps my greatest triumph in that capacity occurred in Uzbekistan, when we became unbelievably thirsty

With Robert Maxwell at Headington Hill Hall in Oxford on the occasion of his 60th birthday. While there, I wrote two autobiographical poems[308] foreshadowing my own 60th birthday later on that year.

while walking in the summer heat in Bukhara. When we came upon a group of Uzbeks sitting under a tree eating cold juicy melons, I persuaded Nakanishi to perform one of his card tricks in return for a melon. Years later I can still taste that melon.

Edward Feigenbaum, the third member of our artificial-intelligence triumvirate, is still a Stanford colleague and friend. He founded Teknowledge, one of the first companies dedicated to industrial applications of knowledge engineering, and invited me to become its initial outside board member.

Donald Glaser, the University of California–Berkeley physics Nobel Prize winner and my friend from an early Pugwash meeting in Sochi in the Soviet Union, was responsible in 1972 for my joining the Board of Cetus Corporation, of which he was a cofounder. We have continued to meet there ever since, in addition to cross-country skiing, musical events, and yearly New Year's Day walks on the beaches at Point Reyes.

With Koji Nakanishi in Sendai in 1964, just after having enjoyed a Japanese bath.

I have already mentioned William Johnson, who brought me to Stanford and who has been the most influential of my Stanford Chemistry Department colleagues. The first Stauffer Chemistry Building was built to our specifications and, to this day, his office is directly above mine. For years, we kept each other's research laboratories in tiptop shape by reciprocal weekly inspections. Neither one of us had the guts to make our own students clean up. Somehow, it was easier to act out the role of tough commissar in somebody else's territory.

I cannot resist bringing up my first encounter in 1957 with Robert Maxwell, owner of Pergamon Press and now also of many other British enterprises. We met together with Sir Robert Robinson, Robert Woodward, Gilbert Stork, and William von Doering at the launching of Pergamon's *Tetrahedron*, whose editorial board we had just joined. Each of us had a famous 100-gold-guinea bet with Maxwell: He didn't believe that *Tetrahedron*'s circulation would exceed that of the *Journal of Organic Chemistry* within 5 years, while we—in our collective cocksure manner—were certain that it would do so, provided we continued as editorial board members. When he triumphantly lost the bet 4 years later, I was the first to get paid: the stunning abstract oil painting from the Gimpel Fils gallery in London is still hanging on my wall. In 1984, our families were joined through the marriage of his daughter Isabel Maxwell to my son Dale, and we now share a grandchild, Alexander Maxwell Djerassi.

The impact of these peers has gone beyond personal and scientific spheres; they have even entered my literary life. Robert Maxwell has already appeared in two of my poems,[308] and several of my friends have been transformed into fictional heroes in my short stories: Gilbert

This diptych commemorates my first and only triumph over Koji Nakanishi's magic. Although always impressed by his performances, I have never been able to figure out any of his more sophisticated sleight-of-hand acts. In February 1990, at the IUPAC Symposium in New Delhi, I served as chairman for Nakanishi's plenary address. The day before, I had bought at an outdoor market two identical ties. After listing Nakanishi's scientific achievements, I ended my introduction by reminding the audience that India seemed to be the only country where Nakanishi had not yet performed his magic. I then fished one of the ties out of my pocket, cut it in half in front of the audience, and put it back into the same pocket. "Let's see him put that together," I challenged Nakanishi before the entire audience, remembering full well how often he had done this with colored ribbons without disclosing the secret of his success. During his talk, in the darkness, I replaced the cut tie with its identical mate, taking care to place the severed one into

---

Stork as Lionel Trippett in "The Toyota Cantos",[309] Koji Nakanishi as Jiko Nishinaka in "Sleight-of-Mind",[310] and Alejandro Zaffaroni as the unnamed chief character in "The Psomophile."[310]

Chemical Fashion in America

It is obvious from what I have written so far that the human component of these decades has been richly rewarding. It is with the science that I

my bag rather than putting it into another one of my pockets. At the end of Nakanishi's lecture, I returned to the podium and asked, "Now let's see whether Nakanishi's magic works in India." I then dared him to reach into my jacket pocket. Nakanishi's response is shown in the second part of the diptych. He proceeded to frisk me in public, looking for some incriminating evidence, but of course was unsuccessful. This was not the end of this magic interlude. I had to leave New Delhi early, but upon my return to Stanford, I received the following note from Nakanishi:

"After the last plenary lecture the organizers asked me to give a 15–20 minute magic performance. I picked Guy Ourisson for the tie trick and did the following. I cut his necktie and put it into my pants pocket, which was turned inside out to show that it was empty. Immediately after putting the cut tie into this empty pocket, I pulled out the crumpled tie from the same pocket and showed that it was in one piece. I will not tell you how I did it, but it was on a principle that I had not worked on before. Thanks for your challenge."

---

am feeling a bit defensive; not with its intrinsic worth, but with a much simpler and cruder index: peer recognition.

America's preeminence in organic chemistry started only around the time of the second World War. Among academic circles, where judgments about prestige are generally made in the context of what is currently fashionable, the American organic chemistry edifice had only two main pillars: physical organic chemistry, with major emphasis on mechanisms of organic reactions; and synthetic organic chemistry, which has become the overriding field in terms of attention and prestige. This includes both "macho" syntheses of exceedingly complicated

With Andre Dreiding, his wife Norma (background), my wife Diane Middlebrook, and the Paul Klee dealer Siegfried Rosengart at the Picasso Museum in Lucerne.

natural products and the development of new synthetic methods and reagents. Natural products research—even in its heyday, when extensive chemical degradation was still required—never carried much prestige or emphasis in the United States. Similarly, even though physical or analytical separation techniques have changed the conduct of organic chemical research more than any other development in the past three to four decades, actual research in these fields by organic chemists has not been accompanied by any particular approbation among American academics.

I can afford to make these statements because, personally, I have little cause for complaint. But let me cite just a few examples, going from west to east. The oldest and most distinguished American organic chemist in the marine natural products field has never received a national American Chemical Society award or been elected to the National Academy of Sciences. Virtually the same can be said about

Four generations of Djerassi males: me, my son Dale, my grandson Alexander, and my father Samuel, each generation separated by 31 years.

Another great reunion with old friends in October 1988. Standing with me at an environmental adobe sculpture (by Ron Fondaw, one of the many visiting artists at the Djerassi Foundation's Artists Colony at SMIP ranch) are E. J. Corey, Jean-Marie Lehn, J. D. Roberts, Duilio Arigoni, and Ronald Breslow. We are on our way to celebrate at a rustic dinner Lehn's Nobel Prize, which had been announced two days earlier.

one of the best organic chemistry professors in the University of California system, whose entire academic career has centered around natural products, or about the leading Midwest practitioner in this chemical subdiscipline. One of the East Coast pioneers in the introduction of mass spectrometry to organic chemistry, especially among alkaloids and peptides, has never been elected to membership in the National Academy of Sciences, in spite of numerous nominations over many years.

Of course, as far as general scientific progress is concerned, it does not matter much that such research may be more favored in Japan or some other country. But that does not change the fact that in the final analysis, peer recognition, especially in one's own country, has an absurdly high value—so much so that I have decided to write about it in a fictional context, where one can be much more open about it. My first novel, *Cantor's Dilemma*,[311] has a great deal to do with a scientist's desire for recognition as well as with related topics such as the mentor–disciple relationship. A second novel in the same genre of

My first literary book-signing event on the occasion of the publication of my novel, Cantor's Dilemma. *Standing in the center of the picture is my stepdaughter, Leah Middlebrook, who was paying a visit from Columbia University.*

science-in-fiction is already underway. My chemical colleagues will probably shudder at the prospect of having to face fiction from the man who seemingly has already published much too much. At least I can offer one consolation: it won't clutter up *Chemical Abstracts* or *Current Contents* (although even that statement is not totally correct[312]).

Coda

coda. something that serves to round out, conclude, or summarize, and that has an interest of its own. (*Webster's Ninth New Collegiate Dictionary*)

In the autumn of 1981, I spent some weeks trekking in western Bhutan near the Tibetan border. The first day was strenuous, starting at 8000 feet and ending with fitful sleep in a tent at an elevation of over 13,000 feet. But from then on, as we crossed ever higher passes on the way to Chomo Lhari, Bhutan's highest and holiest Himalayan peak, I experienced the unforeseen clearing of the mind that I associate with Zen teaching. For many hours, I walked in total solitude in an indescribably majestic and humbling setting. The only man-made sound was my labored breathing and the crunch of my boots against

Trekking in Bhutan near the base camp to Chomo Lhari.

the loose rock. Most days I encountered no other human beings except for the other members of our small party, one of them my son, who was filming a documentary on Bhutan. For once in my life, I had time for day-long reflections in which chemistry had virtually no part.

In the summer of 1985, shortly before I was about to embark on another such period of physical exertion and psychic housecleaning, this time to cross from Tibet into Nepal, Jeffrey Seeman invited me to contribute to this volume. I agreed with the same mental shrug I give when someone invites me to speak on some distant date, because it is simply too far away to worry about. But instead of flying to Asia, on the day of my intended departure I came out of anesthesia after a 5-hour cancer operation with tubes in my nose, arms, and belly. The weeks in the hospital provided another type of reflection, which, originating in the depth of depression, made me come to terms with my own mortality.

Even the healthiest among us know that some day we shall die. Yet over the short and medium course—next month, next year, next decade—we surely think ourselves immortal. I certainly did. Still, even as early as in my middle twenties, I recognized that there was not enough time to do all the things I wanted to accomplish, especially in science. That concern was probably the prime reason for my impatience, for my conviction that time was a precious commodity. I know I

Valery Galkin of the Pacific Institute for Marine Biology in Vladivostok congratulates me for being a "sportist" after canoeing with him in Peter the Great Bay south of Vladivostok near the North Korean border. This trip in September 1989 was striking testimony to the implementation of glasnost; this region of the USSR had been closed to foreigners for more than 50 years. Physically, this experience seemed further evidence of my complete recovery from colon cancer.

did not waste it, but did I spend it wisely and for the right purposes? After my cancer operation, when the morphine-induced cloudiness had lifted, I asked myself whether I would have led a different life if 5 years earlier I had known that I would come down with cancer. It took very little reflection to answer my alter ego in the affirmative. The next question was obvious: knowing what I do now, will I lead a different life? The short answer was "yes." I had the luxury of several more weeks in bed to reflect on the details.

This coda (the editor's idea, not mine) is neither long enough nor the right place to expand on my onco-prompted conclusion. But those weeks in bed confirmed that I was ready for what had first occurred to me fleetingly in the rarefied atmosphere of Bhutan: how exhilarating it might be to lead one more intellectual life, quite different from what I had done during the preceding four decades as a scientist, with all the personal changes that go with such a metamorphosis. The fact that I had the marvelous fortune of marrying, just 2 months before entering the hospital, a spectacularly literate and insightful woman, a true

With my new wife, Diane Middlebrook (Watkins Professor of English at Stanford University) at our wedding party at the SMIP ranch in June 1985.

vocalissima—elegant in word and appearance—helped enormously. It was in this spirit that I finally sat down in front of my word processor to write the preceding account of the portion of my scientific life that had dealt in one way or another with steroids.

That account was written for chemists from the perspective of a chemist. There is much I do not tell for the simple reason that this account would become burdensomely long. For instance, except for a

My son Dale and my daughter Pamela in a picture taken the year before Pamela's suicide.

single sentence in this coda, there is no mention of any of the women who meant so much to me; I have not described my long infatuation with art; I have not dwelled on the greatest tragedy of my life, my daughter's suicide, and how I finally came to terms with that event through the creation of an artists colony, which by now has been the temporary home of more than 500 visual artists, writers, composers, and choreographers. In contrast to my past silence about my private life, since 1985 I started to write about my nonchemical personae. The beginning was a roman a clef, so brutally cathartic as to be unpublishable, but I did permit a few remnants of that cathartic process to float around as poetic trial balloons.[308,313–317] A door has opened, which must have been unlocked or even slightly ajar and which has now swung so wide that on occasion I am still startled by what steps forward.

After a foray, represented by a collection of short stories[310] and a first novel,[311] to test the seriousness of my intellectual commitment to a new literary career, I have decided to proceed concurrently along two paths—one fictional, the other autobiographical. The former, already

A dance performance in 1989 at an open studio event at the Djerassi Foundation's Artist Colony.

well advanced, is in a genre (science in fiction) that offers the opportunity for expressing ideas difficult to discuss in the usual fora open to scientists.

The autobiographical route is provisionally codified under the working title "The Pill, Pygmy Chimps, and Degas's Horse".[318] The first dozen or so memoirs have appeared in literary magazines, under titles like "My Very First Divorce"[319] (an event not found in *Who's Who* or any other biographical entry of mine); "Dear Mrs. Roosevelt"[320]; "White House Enemy"[321] (ostensibly an account of my receiving the National Medal of Science from President Nixon; in actual fact, a record of my second divorce); "Degas's Horse"[322] (the start of my serious art collecting); and "The Quest for Alfred E. Neuman"[323] (describing the length of psychological scars from Hitler's Vienna).

I end this coda by slightly altering one of Cynthia Ozick's great phrases: "These days a hot liquid of remembrance and of imagination lives in the nerve of my joy."

After the dance, with choreographer Victoria Morgan and Eda Holmes, a dancer with the Frankfurt Ballet.

With poet Robert Lowell at an honorary degree ceremony at our alma mater, Kenyon College. Is that where the seed to my poems was planted?

Opera star Tatiana Troyanos autographing my short story "What is Tatiana Troyanos doing in Spartacus's Tent?"[310]

My "very first" wife, Virginia (circa 1945), whom I married before my 20th birthday. She was of enormous help during my graduate school days and the beginning of my professional career. We divorced, childless, after six years of marriage. Forty years later, we discovered that each of us had kept the existence of our first marriage secret for many years from our respective children.

This picture was taken in 1989 shortly after I finished writing my last autobiographical memoir. The question mark above my shoulder may well have a deeper meaning than just referring to the biosynthetic steps on the blackboard. Photo courtesy of Jonas Grushkin.

References

1. Mislow, K. *The Register (Phi Lambda Upsilon)* **1963**, *48(2)*, 46.
2. Djerassi, C.; Zderic, J. A. *J. Am. Chem. Soc.* **1956**, *78*, 2907, 6390.
3. For review, *see* Djerassi, C. In *Festschrift Arthur Stoll*; Birkhauser: Basel, 1957; pp 330–352.
4. Djerassi, C.; Frick, N.; Geller, L. E. *J. Am. Chem. Soc.* **1953**, *75*, 3632.
5. Fitch, W. L.; Dolinger, P. M.; Djerassi, C. *J. Org. Chem.* **1974**, *39*, 2974.
6. Djerassi, C. *Pure Appl. Chem.* **1975**, *41*, 113.
7. *Steroid Reactions: An Outline for Organic Chemists*; Djerassi, C., Ed.; Holden–Day: San Francisco, 1963.
8. Huttrer, C. P.; Djerassi, C.; Beears, W. L.; R. L. Mayer; Scholz, C. R. *J. Am. Chem. Soc.* **1946**, *68*, 1999.
9. Fieser, L. F. *Natural Products Related to Phenanthrene*; Reinhold: New York, 1936.
10. Bachmann, W. E.; Cole, W.; Wilds, A. L. *J. Am. Chem. Soc.* **1940**, *62*, 824.
11. *See* Shapiro, R. H., Chapter 9 in Reference 7, for a review of relevant literature.
12. For leading references, *see* Fieser, L. F.; Fieser, M. *Steroids*; Reinhold: New York, 1959.
13. Inhoffen, H. H. *Angew. Chem.* **1947**, *59*, 207.
14. Wilds, A. L.; Djerassi, C. *J. Am. Chem. Soc.* **1946**, *68*, 1715.
15. Djerassi, C.; Scholz, C. R. *Experientia* **1947**, *3*, 107; idem. *J. Am. Chem. Soc.* **1947**, *69*, 2404; idem. ibid. **1948**, *70*, 417; idem. ibid. **1948**, *70*, 1911; idem. *J. Org. Chem.* **1948**, *13*, 697; idem. ibid. **1949**, *14*, 660.
16. Djerassi, C. *J. Am. Chem. Soc.* **1949**, *71*, 1003.
17. Djerassi, C. *J. Org. Chem.* **1947**, *12*, 823; idem. ibid. **1948**, *13*, 848; idem. *Chem. Rev.* **1948**, *43*, 271.
18. Marker, R. E. *J. Am. Chem. Soc.* **1949**, *71*, 3856.
19. Djerassi, C.; Rosenkranz, G.; Romo, J.; Kaufmann, S.; Pataki, J. *J. Am. Chem. Soc.* **1950**, *72*, 4534.

20. Zderic, J. A.; Bowers, A.; Carpio, H.; Djerassi, C. *J. Am. Chem. Soc.* **1958**, *80*, 2596; Zderic, J. A.; Carpio, H.; Bowers, A.; Djerassi, C. *Steroids* **1963**, *1*, 233.

21. Kaufmann, S.; Pataki, J.; Rosenkranz, G.; Romo, J.; Djerassi, C. *J. Am. Chem. Soc.* **1950**, *72*, 4531.

22. Djerassi, C.; Rosenkranz, G.; Romo, J.; Pataki, J.; Kaufmann, S. *J. Am. Chem. Soc.* **1950**, *72*, 4540.

23. Woodward, R. B.; Singh, G. *J. Am. Chem. Soc.* **1950**, *72*, 494.

24. Sandoval, A.; Miramontes, L.; Rosenkranz, G.; Djerassi, C. *J. Am. Chem. Soc.* **1951**, *73*, 990.

25. Rosenkranz, G.; Pataki, J.; Kaufmann, S.; Berlin, J.; Djerassi, C. *J. Am. Chem. Soc.* **1950**, *72*, 4081.

26. Rosenkranz, G.; Mancera, O.; Gatica, J.; Djerassi, C. *J. Am. Chem. Soc.* **1950**, *72*, 4077.

27. Rosenkranz, G.; Djerassi, C.; Yashin, R.; Pataki, J. *Nature* **1951**, *168*, 28.

28. Fieser, L. F.; Herz, J. E.; Huang, W. *J. Am. Chem. Soc.*, **1951**, *73*, 2397.

29. Chamberlin, E. M.; Ruyle, W. V.; Erickson, A. E.; Chemerda, J. M.; Aliminosa, L. M.; Erickson, R. L.; Sita, G. E.; Tishler, M. *J. Am. Chem. Soc.* **1951**, *73*, 2396.

30. Rosenkranz, G.; Romo, J.; Batres, E.; Djerassi, C. *J. Org. Chem.* **1951**, *16*, 298; Djerassi, C.; Romo, J.; Rosenkranz, G. ibid. **1951**, *16*, 754.

31. Stork, G.; Romo, J.; Rosenkranz, G.; Djerassi, C. *J. Am. Chem. Soc.* **1951**, *73*, 3546; Djerassi, C.; Mancera, O.; Romo, J.; Rosenkranz, G. ibid. **1953**, *75*, 3505.

32. Rosenkranz, G.; Pataki, J.; Djerassi, C. *J. Am. Chem. Soc.* **1951**, *73*, 4055.

33. Chemerda, J. M.; Chamberlin, E. M.; Wilson, E. H.; Tishler, M. *J. Am. Chem. Soc.* **1951**, *73*, 4052.

34. Djerassi, C.; Ringold, H. J.; Rosenkranz, G. *J. Am. Chem. Soc.* **1951**, *73*, 5513.

35. Peterson, D. H.; Murray, H. C. *J. Am. Chem. Soc.* **1952**, *74*, 1871; for a personal account, see Peterson, D. H. *Steroids* **1985**, *45*, 1.

36. Mancera, O.; Zaffaroni, A.; Rubin, B. A.; Sondheimer, F.; Rosenkranz, G.; Djerassi, C. *J. Am. Chem. Soc.* **1952**, *74*, 3711.

37. Haberlandt, L. *Münch. Med. Wochenschr.* 1921, *68*, 1577; Simmer, H. H. *Contraception* 1970, *1*, 3.

38. Ehrenstein, M. *Chem. Rev.* 1948, *42*, 457 and references cited therein.

39. Ehrenstein, M. *J. Org. Chem.* 1944, *9*, 435.

40. Allen, W. M.; Ehrenstein, M. *Science* 1944, *100*, 251.

41. Birch, A. J. *J. Chem. Soc.* 1950, 367.

42. Birch, A. J. *Annu. Rep. Progr. Chem. (Chem. Soc. London)* 1950, *47*, 210.

43. Romo, J.; Rosenkranz, G.; Djerassi, C. *J. Org. Chem.* 1951, *15*, 1289; Sandoval, A.; Rosenkranz, G.; Djerassi, C. *J. Am. Chem. Soc.* 1951, *73*, 2383.

44. Djerassi, C.; Rosenkranz, G.; Iriarte, J.; Berlin, J.; Romo, J. *J. Am. Chem. Soc.* 1951, *73*, 1523. For a different procedure, see Velluz, L.; Muller, G. *Bull. Soc. Chim. Fr.*, 1950, 166.

45. Miramontes, L.; Rosenkranz, G.; Djerassi, C. *J. Am. Chem. Soc.* 1951, *73*, 3540; Djerassi, C.; Miramontes, L.; Rosenkranz, G. ibid. 1953, *75*, 4440.

46. Wilds, A. L.; Nelson, N. A. *J. Am. Chem. Soc.* 1953, *75*, 5366.

47. Tullner, W. W.; Hertz, R. *J. Clin. Endocrinol.* 1952, *12*, 916.

48. Sandoval, A.; Miramontes, L.; Rosenkranz, G.; Djerassi, C.; Sondheimer, F. *J. Am. Chem. Soc.* 1953, *75*, 4117; Sandoval, A.; Thomas, G. H.; Djerassi, C.; Rosenkranz, G.; Sondheimer, F. ibid. 1955, *77*, 148.

49. Zaffaroni, A.; Ringold, H. J.; Rosenkranz, G.; Sondheimer, F.; Thomas, G. H.; Djerassi, C. *J. Am. Chem. Soc.* 1954, *76*, 6210; Zaffaroni, A.; Ringold, H. J.; Rosenkranz, G.; Sondheimer, F.; Thomas, G. H.; Djerassi, C. ibid. 1958, *80*, 6110.

50. Inhoffen, H. H.; Logemann, W.; Hohlweg, W.; Serini, A. *Chem. Ber.* 1938, *71*, 1024.

51. Djerassi, C.; Miramontes, L.; Rosenkranz, G. U.S. Patent 2744 122 (orig. appl. Nov. 22, 1951).

52. Djerassi, C.; Miramontes, L.; Rosenkranz, G. *Abstracts of Papers, Division of Medicinal Chemistry, Milwaukee;* American Chemical Society: Washington, DC, 1952; no. 25, p 18J.

53. Djerassi, C.; Miramontes, L.; Rosenkranz, G.; Sondheimer, F. *J. Am. Chem. Soc.* **1954,** *76,* 4092.

54. Hertz, R.; Tullner, W. W.; Raffelt, E. *Endrocrinology (Baltimore)* **1954,** *54,* 228.

55. Pincus, G.; Chang, M. C.; Hafez, E. S. E.; Zarrow, M. X.; Merrill, A. *Science* **1956,** *124,* 890.

56. Tyler, E. T. Presented at the Annual Meeting of the Pacific Coast Fertility Society, Nov. 1954; *J. Clin. Endocrinol. Metab.* **1955,** *15,* 881; Greenblatt, R. B. ibid. **1956,** *16,* 869; Hertz, R.; White, H. H.; Thomas, L. B. *Proc. Soc. Exp. Biol. Med.* **1956,** *91,* 418.

57. Colton, F. B. U.S. Patent 2 725 389 (orig. appl. Aug. 31, 1953).

58. McGinty, D. A.; Djerassi, C. *Ann. N.Y. Acad. Sci.* **1958,** *71,* 500.

59. Djerassi, C. The Politics of Contraception; W. H. Freeman: San Francisco, 1981; pp 227–256.

60. Smith, H.; et al. *J. Chem. Soc.* **1964,** 4472.

61. Barton, D. H. R. *J. Chem. Soc.* **1945,** 813, and later papers.

62. Klyne, W. In *Determination of Organic Structures by Physical Methods;* Braude, E. A.; Nachod, F. C., Eds.; Academic Press: New York, 1955.

63. Inter al. Djerassi, C.; Foltz, E. W.; Lippman, A. E. *J. Am. Chem. Soc.* **1955,** *77,* 4354; Djerassi, C.; Closson, W.; Lippman, A. E. ibid. **1956,** *78,* 3163.

64. Djerassi, C.; Riniker, R.; Riniker, B. *J. Am. Chem. Soc.* **1956,** *78,* 6362.

65. Djerassi, C.; Halpern, O.; Halpern, V.; Riniker, B. *J. Am. Chem. Soc.* **1958,** *80,* 4001.

66. Ruzicka, L. *Experientia* **1953,** *9,* 357.

67. Djerassi, C.; Rittel, W.; Nussbaum, A. L.; Donovan, F. W.; Herran, J. *J. Am. Chem. Soc.* **1954,** *76,* 6410.

68. Djerassi, C.; Rittel, W. *J. Am. Chem. Soc.* **1957,** *79,* 3528.

69. Djerassi, C.; Burstein, S. *J. Am. Chem. Soc.* **1958,** *80,* 2593.

70. Djerassi, C.; Osiecki, J.; Riniker, R.; Riniker, B. *J. Am. Chem. Soc.* **1958,** *80,* 1216.

71. Djerassi, C.; Klyne, W. *J. Am. Chem. Soc.* **1957,** *79,* 1506.

72. Moffitt, W.; Woodward, R. B.; Moscowitz, A.; Klyne, W.; Djerassi, C. *J. Am. Chem. Soc.* **1961,** *83,* 4013.

73. *Current Contents* Oct. 18, 1982, 42, 22.

74. Lightner, D. A.; Bouman, T. D.; Wijekoon, W. M. D.; Hansen, A. E. *J. Am. Chem. Soc.* **1986**, *108*, 4484, and earlier papers.

75. Djerassi, C.; Geller, L. E. *Tetrahedron* **1958**, *3*, 319; Djerassi, C.; Geller, L. E.; Eisenbraun, E. J. *J. Org. Chem.* **1960**, *25*, 1.

76. Allinger, J.; Allinger, N. L. *Tetrahedron* **1958**, *5*, 64.

77. Iriarte, J.; Djerassi, C.; Ringold, H. J. *J. Am. Chem. Soc.* **1959**, *81*, 436; Villotti, R.; Djerassi, C.; Ringold, H. J. ibid. **1959**, *81*, 4566.

78. Ringold, H. J.; Mancera, O.; Djerassi, C.; Bowers, A.; Batres, E.; Martinez, H.; Necoechea, E.; Edwards, J.; Velasco, M.; Casas Campillo, C.; Dorfman, R. I. *J. Am. Chem. Soc.* **1958**, *80*, 6464; Mills, J. S.; Bowers, A.; Casas Campillo, C.; Djerassi, C.; Ringold, H. J. ibid. **1959**, *81*, 1264; Edwards, J. A.; Ringold, H. J.; Djerassi, C. ibid. **1959**, *81*, 3156; Zderic, J. A.; Carpio, H.; Djerassi, C. *J. Org. Chem.* **1959**, *24*, 909; Edwards, J. A.; Ringold, H. J.; Djerassi, C. *J. Am. Chem. Soc.* **1960**, *82*, 2318; Mills, J. S.; Bowers, A.; Djerassi, C.; Ringold, H. J. ibid. **1960**, *82*, 3399.

79. Mills, J. S.; Ringold, H. J.; Djerassi, C. *J. Am. Chem. Soc.* **1958**, *80*, 6118; Knox, L.; Zderic, J. A.; Perez Ruelas, J.; Djerassi, C.; Ringold, H. J. ibid. **1960**, *82*, 1230; Mills, J. S.; Candiani, O.; Djerassi, C. *J. Org. Chem.* **1960**, *25*, 1056.

80. Djerassi, C.; Fornaguera, I.; Mancera, O. *J. Am. Chem. Soc.* **1959**, *81*, 2383; Djerassi, C.; Finch, N.; Mauli, R. ibid. **1959**, *81*, 4997; Mauli, R.; Ringold, H. J.; Djerassi, C. ibid. **1960**, *82*, 5494; Villotti, R.; Ringold, H. J.; Djerassi, C. ibid. **1960**, *82*, 5693.

81. Djerassi, C. *Optical Rotatory Dispersion: Applications to Organic Chemistry*; McGraw–Hill: New York, 1960.

82. Velluz, L.; Legrand, M. *Angew. Chem.* **1961**, *73*, 603; Legrand, M.; Mathieu, J. *Bull. Soc. Chim. Fr.* **1961**, 1679.

83. Badoz, J.; Billardon, M.; Mathieu, J.-P. *Compt. Rend.* **1960**, *251*, 1477; Grosjean, M.; Legrand, M. ibid. **1960**, *251*, 2150.

84. Djerassi, C.; Wolf, H.; Bunnenberg, E. *J. Am. Chem. Soc.* **1963**, *85*, 324.

85. Mislow, K.; Glass, M. A. W.; O'Brien, R. E.; Rutkin, P.; Steinberg, D. M.; Weiss, J.; Djerassi, C. *J. Am. Chem. Soc.* **1962**, *84*, 1455.

86. Mislow, K.; Glass, M. A. W.; Moscowitz, A.; Djerassi, C. *J. Am. Chem. Soc.* **1961**, *83*, 2771.

87. Moscowitz, A.; Mislow, K.; Glass, M. A. W.; Djerassi, C. *J. Am. Chem. Soc.* **1962**, *84*, 1945.

88. Mislow, K.; Bunnenberg, E.; Records, R.; Wellman, K.; Djerassi, C. *J. Am. Chem. Soc.* **1963**, *85*, 1342.

89. Sjöberg, B.; Fredga, A.; Djerassi, C. *J. Am. Chem. Soc.* **1959**, *81*, 5002; Sjöberg, B.; Cram, D. J.; Wolf, L.; Djerassi, C. *Acta Chem. Scand.* **1962**, *16*, 1079.

90. Djerassi, C.; Wolf, H.; Bunnenberg, E. *J. Am. Chem. Soc.* **1962**, *84*, 4552.

91. Djerassi, C.; Undheim, K. *J. Am. Chem. Soc.* **1960**, *82*, 5755; Djerassi, C.; Undheim, K.; Weidler, A.-M. *Acta Chem. Scand.* **1962**, *16*, 1147.

92. Djerassi, C.; Lund, E.; Bunnenberg, E.; Sjöberg, B. *J. Am. Chem. Soc.* **1961**, *83*, 2307.

93. Djerassi, C.; Harrison, I. T.; Zagneetko, O.; Nussbaum, A. L. *J. Org. Chem.* **1962**, *27*, 1173; Djerassi, C.; Wolf, H.; Bunnenberg, E. *J. Am. Chem. Soc.* **1963**, *85*, 2835.

94. Djerassi, C. *Proc. Chem. Soc. London* **1964**, 314.

95. Wellman, K. M.; Bunnenberg, E.; Djerassi, C. *J. Am. Chem. Soc.* **1963**, *85*, 1870; Wellman, K. M.; Records, R.; Bunnenberg, E.; Djerassi, C. ibid. **1964**, *86*, 492.

96. Moscowitz, A.; Wellman, K. M.; Djerassi, C. *J. Am. Chem. Soc.* **1963**, *85*, 3515.

97. Moscowitz, A.; Wellman, K. M.; Djerassi, C. *Proc. Natl. Acad. Sci. U.S.A.* **1963**, *50*, 799.

98. Wellman, K. M.; Djerassi, C. *J. Am. Chem. Soc.* **1965**, *87*, 60.

99. Duax, W. L.; Griffin, J. F.; Rohrer, D. C. *J. Am. Chem. Soc.* **1981**, *103*, 6705.

100. Briggs, W. S.; Djerassi, C. *Tetrahedron* **1965**, *21*, 3455.

101. Eliel, E. L. *J. Am. Chem. Soc.* **1949**, *71*, 3970.

102. Verbit, L. *Progr. Phys. Org. Chem.* **1970**, *7*, 51.

103. Djerassi, C.; Tursch, B. *J. Am. Chem. Soc.* **1961**, *83*, 4609.

104. Djerassi, C.; Krakower, G. W. *J. Am. Chem. Soc.* **1959**, *81*, 237.

105. Simek, J. W.; Mattern, D. C.; Djerassi, C. *Tetrahedron Lett.* **1975**, 3671.

106. Sundararaman, P.; Djerassi, C. *Tetrahedron Lett.* **1978**, 2457; Errata, ibid. **1979**, 4120.

107. Sundararaman, P.; Barth, G.; Djerassi, C. *J. Org. Chem.* **1980**, *45*, 5231.

108. Lee, S.-F.; Edgar, M. T.; Pak, C. S.; Barth, G.; Djerassi, C. *J. Am. Chem. Soc.* **1980**, *102*, 4784.

109. Pak, C. S.; Djerassi, C. *Tetrahedron Lett.* **1978**, 4377.

110. Djerassi, C.; VanAntwerp, C. L.; Sundararaman, P. *Tetrahedron Lett.* **1978**, 535.

111. Lee, S.-F.; Barth, G.; Kieslich, K.; Djerassi, C. *J. Am. Chem. Soc.* **1978**, *100*, 3965; Lee, S.-F.; Barth, G.; Djerassi, C. ibid. **1981**, *103*, 295.

112. Barth, G.; Djerassi, C. *Tetrahedron* **1981**, *37*, 4123.

113. Lightner, D. A.; Chang, T. C.; Horwitz, J. *Tetrahedron Lett.* **1977**, 3019; Errata, ibid. **1978**, 696.

114. Numan, H.; Wynberg, H. *J. Org. Chem.* **1978**, *43*, 2232.

115. Sing, Y. L.; Numan, H.; Wynberg, H.; Djerassi, C. *J. Am. Chem. Soc.* **1979**, *101*, 5155; Errata, ibid. **1979**, *101*, 7439.

116. Lightner, D. A.; Gawronski, J. K.; Bouman, T. D. *J. Am. Chem. Soc.* **1980**, *102*, 1983.

117. Djerassi, C.; Knight, J. C.; Brockmann, H. *Chem. Ber.* **1964**, *97*, 3118, and references cited therein.

118. Djerassi, C.; Mills, J. S.; Villotti, R. *J. Am. Chem. Soc.* **1958**, *80*, 1005; Djerassi, C.; Krakower, G. W., Lemin, A. J., Liu, L. H.; Mills, J. S.; Villoti, R. ibid. **1958**, *80*, 6284.

119. De Mayo, P.; Reed, R. I. *Chem. Ind. (London)* **1956**, 1481.

120. Nilsson, M.; Ryhage, R.; von Sydow, E. *Acta Chem. Scand.* **1957**, *11*, 634; Ryhage, R.; Stenhagen, E. *Arkiv. Kemi.* **1959**, *15*, 291; Stenhagen, E. *Z. Anal. Chem.* **1961**, *181*, 462.

121. Biemann, K.; Spiteller-Friedmann, M.; Spiteller, G. *Tetrahedron Lett.* **1961**, 458; Biemann, K. *Mass Spectrometry;* McGraw–Hill: New York, 1962.

122. Budzikiewicz, H.; Djerassi, C. *J. Am. Chem. Soc.* **1962**, *84*, 1430.

123. Budzikiewicz, H.; Djerassi, C.; Williams, D. H. *Interpretation of Mass Spectra of Organic Compounds;* Holden–Day: San Francisco, 1964.

124. Budzikiewicz, H.; Djerassi, C.; Williams, D. H. *Structure Elucidation of Natural Products by Mass Spectrometry*; Holden–Day: San Francisco, 1964; Vol. 1: Alkaloids.

125. Budzikiewicz, H.; Djerassi, C.; Williams, D. H. *Structure Elucidation of Natural Products by Mass Spectrometry*; Holden–Day: San Francisco, 1964; Vol. 2: Steroids, Terpenoids, Sugars, and Miscellaneous Classes.

126. Budzikiewicz, H.; Djerassi, C.; Williams, D. H. *Mass Spectrometry of Organic Compounds*; Holden–Day: San Francisco, 1967.

127. Djerassi, C. *Pure Appl. Chem.* **1963**, *6*, 575.

128. Djerassi, C.; Wilson, J. M.; Budzikiewicz, H.; Chamberlin, J. W. *J. Am. Chem. Soc.* **1962**, *84*, 4544.

129. Djerassi, C.; Budzikiewicz, H.; Wilson, J. M. *Tetrahedron Lett.* **1962**, 263.

130. Budzikiewicz, H.; Wilson, J. M.; Djerassi, C. *J. Am. Chem. Soc.* **1963**, *85*, 3688.

131. Djerassi, C.; Budzikiewicz, H.; Wilson, J. M.; Gosset, J.; LeMen, J.; Janot, M.-M. *Tetrahedron Lett.* **1962**, 235.

132. Plat, M.; LeMen, J.; Janot, M.-M.; Wilson, J. M.; Budzikiewicz, H.; Durham, L. J.; Nakagawa, Y.; Djerassi, C. *Tetrahedron Lett.* **1962**, 271.

133. Budzikiewicz, H.; Brauman, J. I.; Djerassi, C. *Tetrahedron* **1965**, *21*, 1855.

134. Hammerum, S.; Djerassi, C. *J. Am. Chem. Soc.* **1973**, *95*, 5806.

135. Dixon, J. S.; Midgley, I.; Djerassi, C. *J. Am. Chem. Soc.* **1977**, *99*, 3432.

136. Chapter 2 in reference 124.

137. Djerassi, C. *Proceedings of the Second International Congress on Hormonal Steroids*, Excerpta Medica Foundation: Amsterdam, 1967, pp 261–268.

138. Voelter, W.; Djerassi, C. *Chem. Ber.* **1968**, *101*, 58.

139. Fischer, M.; Pelah, Z.; Williams, D. H.; Djerassi, C. *Chem. Ber.* **1965**, *98*, 3236.

140. Midgley, I.; Djerassi, C. *Tetrahedron Lett.* **1972**, 4673.

141. Taylor, E. J.; Djerassi, C. *J. Am. Chem. Soc.* **1976**, *98*, 2275.

142. Grant, B.; Djerassi, C. *J. Org. Chem.* **1974**, *39*, 968.

143. Djerassi, C. *Pure Appl. Chem.* **1964**, *9*, 159.

144. Djerassi, C. In *Advances in Mass Spectrometry*; Kendrick, C. E., Ed.; Institute of Petroleum: London, 1968; Vol. 4, pp 199–210.

145. Djerassi, C. *Pure Appl. Chem.* **1970**, *21*, 205.

146. Djerassi, C. *Pure Appl. Chem.* **1978**, *50*, 171.

147. For review, see Chapter 3 in reference 126.

148. Beard, C.; Wilson, J. M.; Budzikiewicz, H.; Djerassi, C. *J. Am. Chem. Soc.* **1964**, *86*, 269.

149. Williams, D. H.; Wilson, J. M.; Budzikiewicz, H.; Djerassi, C. *J. Am. Chem. Soc.* **1963**, *85*, 2091.

150. Williams, D. H.; Djerassi, C. *Steroids* **1964**, *3*, 259.

151. Djerassi, C.; von Mutzenbecher, G.; Fajkos, J.; Williams, D. H.; Budzikiewicz, H. *J. Am. Chem. Soc.* **1965**, *87*, 817.

152. Shapiro, R. H.; Wilson, J. M.; Djerassi, C. *Steroids* **1963**, *1*, 1; Shapiro, R. H.; Djerassi, C. *J. Am. Chem. Soc.* **1964**, *86*, 2825.

153. Brown, F. J.; Djerassi, C. *J. Am. Chem. Soc.* **1980**, *102*, 807; Brown, F. J.; Djerassi, C. *J. Org. Chem.* **1981**, *46*, 955.

154. See Chapter 6 in reference 126.

155. Audier, H.; Diara, A.; de J. Durazo, M.; Fetizon, M.; Foy, P.; Vetter, W. *Bull. Soc. Chim. Fr.* **1963**, 2827.

156. von Mutzenbecher, G.; Pelah, Z.; Williams, D. H.; Budzikiewicz, H.; Djerassi, C. *Steroids* **1963**, *2*, 475.

157. Pelah, Z.; Williams, D. H.; Budzikiewicz, H.; Djerassi, C. *J. Am. Chem. Soc.* **1964**, *86*, 3722.

158. For leading references, see Chapter 21 in reference 125.

159. Zaretskii, Z. V. I. *Mass Spectrometry of Steroids*; Wiley: New York, 1976.

160. Spiteller, G.; Spiteller-Friedmann, M. *Liebigs Ann. Chem.* **1965**, *690*, 1.

161. Reed, R. I. *J. Chem. Soc.*, **1958**, 3432.

162. Friedland, S. S.; Lane, G. H.; Longman, R. T.; Train, K. E.; O'Neal, M. J. *Anal. Chem.* **1959**, *31*, 169.

163. Ryhage, R.; Stenhagen, E. *J. Lipid Res.* **1960**, *1*, 361.

164. Tökés, L.; Jones, G.; Djerassi, C. *J. Am. Chem. Soc.* **1968**, *90*, 5465.

165. Patterson, D. G.; Haley, M. J.; Midgley, I.; Djerassi, C. *Org. Mass Spectrom.* **1984**, *19*, 531.

166. Spiteller-Friedmann, M.; Eggers, S.; Spiteller, G. *Monatsh. Chem.* **1964**, *95*, 1740.

167. Eadon, G.; Popov, S.; Djerassi, C. *J. Am. Chem. Soc.* **1972**, *94*, 1282.

168. Partridge, L. G.; Midgley, I.; Djerassi, C. *J. Am. Chem. Soc.* **1977**, *99*, 7686.

169. Wyllie, S. G.; Djerassi, C. *J. Org. Chem.* **1968**, *33*, 305.

170. Massey, I. J.; Djerassi, C. *J. Org. Chem.* **1979**, *44*, 2448.

171. Theobald, N.; Wells, R. J.; Djerassi, C. *J. Am. Chem. Soc.* **1978**, *100*, 7677.

172. Brown, P.; Djerassi, C. *J. Am. Chem. Soc.* **1966**, *88*, 2469.

173. Komitsky, F., Jr.; Gurst, J. E.; Djerassi, C. *J. Am. Chem. Soc.* **1965**, *87*, 1398; Harris, R. L. N.; Komitsky, F., Jr.; Djerassi, C. ibid. **1967**, *89*, 4765.

174. Jaeger, D. A.; Tomer, K. B.; Woodgate, P. D.; Gebreyesus, T.; Djerassi, C.; Shapiro, R. H. *Org. Mass Spectrom.* **1974**, *9*, 551.

175. Green, M. M.; Djerassi, C. *J. Am. Chem. Soc.* **1967**, *89*, 5190.

176. Biemann, K. *Pure Appl. Chem.* **1964**, *9*, 95.

177. Brown, P.; Djerassi, C. *Angew. Chem. Int. Ed. Engl.* **1967**, *6*, 477.

178. *Current Contents* **Aug. 23, 1982**, *34*, 18.

179. Shashoua, V. *J. Am. Chem. Soc.* **1964**, *86*, 2109.

180. Briat, B.; Billardon, M.; Badoz, J. *Compt. Rend.* **1963**, *256*, 3440; Briat, B. ibid. **1964**, *258*, 2788.

181. Buckingham, A. D.; Stephens, P. J. *Ann. Rev. Phys. Chem.* **1966**, *17*, 399.

182. Schooley, D. A.; Bunnenberg, E.; Djerassi, C. *Proc. Natl. Acad. Sci. U.S.A.* **1965**, *53*, 579; idem. ibid. **1966**, *56*, 1377.

183. Briat, B.; Schooley, D. A.; Records, R.; Bunnenberg, E.; Djerassi, C. *J. Am. Chem. Soc.* **1967**, *89*, 6170.

184. Briat, B.; Djerassi, C. *Nature* **1968**, *217*, 918.

185. McCaffery, A. J.; Henning, G. N.; Schatz, P. N.; Ritchie, A. B.; Perzanowski, H. P.; Rodig, O. R.; Norvelle, A. W.; Stephens, P. J. *Chem. Commun.* **1966**, 250; Winkler, J. ibid. **1966**, 948.

186. Barth, G.; Bunnenberg, E.; Djerassi, C. *Chem. Commun.* **1969**, 1246; Barth, G.; Bunnenberg, E.; Djerassi, C.; Elder, D.; Records, R. *Symp. Faraday Soc.* **1969**, No. 3, 49.

187. Seamans, L.; Moscowitz, A.; Barth, G.; Bunnenberg, E.; Djerassi, C. *J. Am. Chem. Soc.* **1972**, *94*, 6464.

188. Seamans, L.; Moscowitz, A.; Linder, R. E.; Morrill, K.; Dixon, J. S.; Barth, G.; Bunnenberg, E.; Djerassi, C. *J. Am. Chem. Soc.* **1977**, *99*, 724; Linder, R. E.; Morrill, K.; Dixon, J. S.; Barth, G.; Bunnenberg, E.; Djerassi, C.; Seamans, L.; Moscowitz, A. ibid. **1977**, *99*, 727.

189. For an early review, see Djerassi, C.; Bunnenberg, E.; Elder, D. *Pure Appl. Chem.* **1971**, *25*, 57.

190. Barth, G.; Records, R.; Bunnenberg, E.; Djerassi, C.; Voelter, W. *J. Am. Chem. Soc.* **1971**, *93*, 2545; idem. ibid. **1972**, *94*, 1293; Barth, G.; Bunnenberg, E.; Djerassi, C. *Anal. Biochem.* **1972**, *48*, 471; Barth, G.; Linder, R. E.; Bunnenberg, E.; Djerassi, C. *Liebigs Ann. Chem.* **1974**, 990.

191. Dolinger, P. M.; Kielczewski, M.; Trudell, J. R.; Barth, G.; Linder, R. E.; Bunnenberg, E.; Djerassi, C. *Proc. Natl. Acad. Sci. U.S.A.* **1974**, *71*, 399; Dawson, J. H.; Dolinger, P. M.; Trudell, J. R.; Bunnenberg, E.; Barth, G.; Linder, R. E.; Djerassi, C. ibid. **1974**, *71*, 4594; Dawson, J. H.; Trudell, J. R.; Barth, G.; Linder, R. E.; Bunnenberg, E.; Djerassi, C.; Gouterman, M.; Connell, C. R.; Sayer, P. *J. Am. Chem. Soc.* **1977**, *99*, 641; Dawson, J. H.; Trudell, J. R.; Linder, R. E.; Barth, G.; Bunnenberg, E.; Djerassi, C. *Biochemistry* **1978**, *17*, 33.

192. Collman, J. P.; Sorrell, T. N.; Dawson, J. H.; Trudell, J. R.; Bunnenberg, E.; Djerassi, C. *Proc. Natl. Acad. Sci. U.S.A.* **1976**, *73*, 6; Dawson, J. H.; Holm, R. H.; Trudell, J. R.; Barth, G.; Linder, R. E.; Bunnenberg, E.; Djerassi, C.; Tang, S. C. *J. Am. Chem. Soc.* **1976**, *98*, 3707.

193. Dawson, J. H.; Trudell, J. R.; Barth, G.; Linder, R. E.; Bunnenberg, E.; Djerassi, C.; Chiang, R.; Hager, L. P. *J. Am. Chem. Soc.* **1976**, *98*, 3709.

194. Barth, G.; Linder, R. E.; Bunnenberg, E.; Djerassi, C. *Ann. N.Y. Acad. Sci.* **1973**, *206*, 223.

195. Barth, G.; Linder, R. E.; Bunnenberg, E.; Djerassi, C. *J. Chem. Soc. Perkin Trans.* 2 **1974**, 696.

196. Barth, G.; Linder, R. E.; Bunnenberg, E.; Djerassi, C.; Seamans, L.; Moscowitz, A. *J. Chem. Soc., Perkin Trans.* 2 **1974**, 1706; Linder, R. E.; Barth, G.; Bunnenberg, E.; Djerassi, C.; Seamans, L.; Moscowitz, A. ibid. **1974**, 1712; Barth, G.; Linder, R. E.; Waespe-Sarcevic, N.; Bunnenberg, E.; Djerassi, C.; Aronowitz, Y. J.; Gouterman, M. ibid. **1977**, 337.

197. Keegan, J. D.; Stolzenberg, A. M.; Lu, Y.-C.; Linder, R. E.; Barth, G.; Bunnenberg, E.; Djerassi, C.; Moscowitz, A. *J. Am. Chem. Soc.* **1981**, *103*, 3201; Keegan, J. D.; Stolzenberg, A. M.; Lu, Y.-C.; Linder, R. E.; Barth, G.; Moscowitz, A.; Bunnenberg, E.; Djerassi, C. ibid. **1982**, *104*, 4305; Djerassi, C.; Lu, Y.; Waleh, A.; Shu, A. Y. L.; Goldbeck, R. A.; Kehres, L. A.; Crandell, C. W.; Wee, A. G. H.; Knierzinger, A.; Gaete-Homes, R.; Loew, G.; Clezy, P. S.; Bunnenberg, E. ibid. **1984**, *106*, 4241.

198. Keegan, J. D.; Stolzenberg, A. M.; Lu, Y.-C.; Linder, R. E.; Barth, G.; Moscowitz, A.; Bunnenberg, E.; Djerassi, C. *J. Am. Chem. Soc.* **1982**, *104*, 4317.

199. For relevant reviews, see Michl, J. *Pure Appl. Chem.* **1980**, *52*, 1549; idem. *Tetrahedron* **1984**, *40*, 3845.

200. Lu, Y.; Shu, A. Y. L.; Knierzinger, A.; Clezy, P. S.; Bunnenberg, E.; Djerassi, C. *Tetrahedron Lett.* **1983**, *24*, 2433; Goldbeck, R. A.; Tolf, B.-R.; Wee, A. G. H.; Shu, A. Y. L.; Records, R.; Bunnenberg, E.; Djerassi, C. *J. Am. Chem. Soc.* **1986**, *108*, 6449.

201. Schlabach, M.; Wehrle, B.; Limbach, H.-H.; Bunnenberg, E.; Knierzinger, A.; Shu, A. Y. L.; Tolf, B.-R.; Djerassi, C. *J. Am. Chem. Soc.* **1986**, *108*, 3856.

202. Wee, A. G. H.; Shu, A. Y. L.; Bunnenberg, E.; Djerassi, C. *J. Org. Chem.* **1984**, *49*, 3327; Goldbeck, R. A.; Tolf, B.-R.; Bunnenberg, E.; Djerassi, C. *J. Am. Chem. Soc.* **1987**, *109*, 28.

203. Jackson, A. H.; Kenner, G. W.; Smith, K. M.; Aplin, R. T.; Budzikiewicz, H.; Djerassi, C. *Tetrahedron* **1965**, *21*, 2913.

204. Jiang, X.-Y.; Szente, A. W.; Tolf, B.-R.; Kehres, L. A.; Bunnenberg, E.; Djerassi, C. *Tetrahedron Lett.* **1984**, *25*, 4083; Tolf, B.-R.; Jiang, X.-Y.; Wegmann-Szente, A.; Kehres, L. A.; Bunnenberg, E.; Djerassi, C. *J. Am. Chem. Soc.* **1986**, *108*, 1363.

205. For leading references, see Shaw, G. J.; Eglinton, G.; Quirke, J. M. E. *Anal. Chem.* **1981**, *53*, 2014; Sundararaman, P.; Gallegos, E. J.; Baker, E. W.; Stayback, J. R. B.; Johnston, M. R. ibid. **1984**, *56*, 2552.

206. Lederberg, J. *Proc. Natl. Acad. Sci. U.S.A.* **1965**, *53*, 134; idem. Report No. CR–57029 & STAR No. N65–13158 (1964), Report No. CR–68898 & STAR No. N66–14074 (1965), Report No. CR–68899 & STAR No. N66–14075 (1966); National Aeronautics and Space Administration: Washington, DC.

207. Lindsay, R.; Buchanan, B. G.; Feigenbaum, E. A.; Lederberg, J. *Applications of Artificial Intelligence to Organic Chemistry;* McGraw–Hill: New York, 1980.

208. Djerassi, C.; Smith, D. H.; Crandell, C. W.; Gray, N. A. B.; Nourse, J. G.; Lindley, M. R. *Pure Appl. Chem.* **1982**, *54*, 2425.

209. Buchs, A.; Delfino, A. B.; Duffield, A. M.; Djerassi, C.; Buchanan, B. G.; Feigenbaum, E. A.; Lederberg, J. *Helv. Chim. Acta* **1970**, *53*, 1394; Buchs, A.; Delfino, A. B.; Djerassi, C.; Duffield, A. M.; Buchanan, B. G.; Feigenbaum, E. A.; Lederberg, J.; Schroll, G.; Sutherland, G. L. *Advances in Mass Spectrometry* **1971**, *5*, 314.

210. Lederberg, J.; Sutherland, G. L.; Buchanan, B. G.; Feigenbaum, E. A.; Robertson, A. V.; Duffield, A. M.; Djerassi, C. *J. Am. Chem. Soc.* **1969**, *91*, 2973.

211. Sheikh, Y. M.; Buchs, A.; Delfino, A. B.; Schroll, G.; Duffield, A. M.; Djerassi, C.; Buchanan, B. G.; Sutherland, G. L.; Feigenbaum, E. A.; Lederberg, J. *Org. Mass Spectrom.* **1970**, *4*, 493.

212. Duffield, A. M.; Robertson, A. V.; Djerassi, C.; Buchanan, B. G.; Sutherland, G. L.; Feigenbaum, E. A.; Lederberg, J. *J. Am. Chem. Soc.* **1969**, *91*, 2977.

213. Schroll, G.; Duffield, A. M.; Djerassi, C.; Buchanan, B. G.; Sutherland, G. L.; Feigenbaum, E. A.; Lederberg, J. *J. Am. Chem. Soc.* **1969**, *91*, 7440.

214. Buchs, A.; Duffield, A. M.; Djerassi, C.; Schroll, G.; Delfino, A. B.; Buchanan, B. G.; Sutherland, G. L.; Feigenbaum, E. A.; Lederberg, J. *J. Am. Chem. Soc.* **1970**, *92*, 6831.

215. Smith, D. H., Buchanan, B. G.; Engelmore, R. S.; Duffield, A. M.; Yeo, A.; Feigenbaum, E. A.; Lederberg, J.; Djerassi, C. *J. Am. Chem. Soc.* **1972**, *94*, 5962; Smith, D. H.; Buchanan, B. G.; Engelmore, R. S.; Adlercreutz, H.; Djerassi, C. ibid. **1973**, *95*, 6078; Smith, D. H.; Buchanan, B. G.; White, W. C.; Feigenbaum, E. A.; Lederberg, J.; Djerassi, C. *Tetrahedron* **1973**, *29*, 3117.

216. Carhart, R. E.; Djerassi, C. *J. Chem. Soc. Perkin Trans. 2* **1973**, 1753.

217. Eggert, H.; Djerassi, C. *J. Org. Chem.* **1973**, *38*, 3788; idem. *Tetrahedron Lett.* **1975**, 3635; Eggert, H.; VanAntwerp, C. L.; Bhacca, N. S.; Djerassi, C. *J. Org. Chem.* **1976**, *41*, 71; VanAntwerp, C.; Eggert, H.; Meakins, G. D.; Miners, J. O.; Djerassi, C. ibid. **1977**, *42*, 789; Eggert, H.; Djerassi, C. ibid. **1981**, *46*, 5399.

218. Carhart, R. E.; Smith, D. H.; Brown, H.; Djerassi, C. *J. Am. Chem. Soc.* **1975**, *97*, 5755.

219. Djerassi, C.; Smith, D. H.; Varkony, T. H. *Naturwissenschaften* **1979**, *66*, 9.

220. Carhart, R. E.; Smith, D. H.; Gray, N. A. B.; Nourse, J. G.; Djerassi, C. *J. Org. Chem.* **1981**, *46*, 1708.

221. Nourse, J. G.; Carhart, R. E.; Smith, D. H.; Djerassi, C. *J. Am. Chem. Soc.* **1979**, *101*, 1216.

222. Nourse, J. G.; Smith, D. H.; Carhart, R. E.; Djerassi, C. *J. Am. Chem. Soc.* **1980**, *102*, 6289.

223. Gray, N. A. B.; Crandell, C. W.; Nourse, J. G.; Smith, D. H.; Dageforde, M. L.; Djerassi, C. *J. Org. Chem.* **1981**, *46*, 703.

224. Gray, N. A. B.; Nourse, J. G.; Crandell, C. W.; Smith, D. H.; Djerassi, C. *Org. Magn. Reson.* **1981**, *15*, 375.

225. Ohno, N.; Mabry, T. J.; Zabel, V.; Watson, W. H. *Phytochemistry*, **1979**, *18*, 1687.

226. Egli, H.; Smith, D. H.; Djerassi, C. *Helv. Chim. Acta* **1982**, *65*, 1898.

227. Lindley, M. R.; Shoolery, J. N.; Smith, D. H.; Djerassi, C. *Org. Magn. Reson.* **1983**, *21*, 405.

228. Karliner, J.; Budzikiewicz, H.; Djerassi, C. *J. Org. Chem.* **1966**, *31*, 710; Karliner, J.; Djerassi, C. ibid. **1966**, *31*, 1945.

229. Djerassi, C. *Proceedings of the Second International Congress on Hormonal Steroids*; Excerpta Medica Foundation: Amsterdam, 1967; pp 3–15.

230. Bergmann, W.; McLean, M. J.; Lester, D. J. *J. Org. Chem.* **1943**, *8*, 271.

231. Hale, R. L.; Leclercq, J.; Tursch, B.; Djerassi, C.; Gross, R. A., Jr.; Weinheimer, A. J.; Gupta, K.; Scheuer, P. J. *J. Am. Chem. Soc.* **1970**, *92*, 2179.

232. Ling, N. C.; Hale, R. L.; Djerassi, C. *J. Am. Chem. Soc.*, **1970**, *92*, 5281.

233. Itoh, T.; Sica, D.; Djerassi, C. *J. Chem. Soc. Perkin Trans. 1* **1983**, 147.

234. Ciereszko, L. S.; Johnson, M. A.; Schmidt, R. W.; Koons, C. B. *Comp. Biochem. Physiol.* **1968**, *24*, 899.

235. Minale, L.; Sodano, G. In *Marine Natural Products Chemistry;* Faulkner, D. J.; Fenical, W. H., Eds.; Plenum Press: New York, 1977; pp 87–109.

236. Schmitz, F. J. In *Marine Natural Products;* Scheuer, P. J., Ed.; Academic Press: New York, 1978; Vol. 1, Chapter 5.

237. Goad, L. J. In *Marine Natural Products;* Scheuer, P. J., Ed.; Academic Press: New York, 1978; Vol. 2, Chapter 2.

238. Djerassi, C.; Theobald, N.; Kokke, W. C. M. C.; Pak, C. S.; Carlson, R. M. K. *Pure Appl. Chem.* **1979**, *51*, 1815.

239. Djerassi, C. *Pure Appl. Chem.* **1981**, *53*, 873.

240. Djerassi, C. In *Proceedings of the Alfred Benzon Symposium 20;* Krogsgaard-Larsen, P.; Brogger Christensen, S.; Kofod, H., Eds.; Munksgaard: Copenhagen, 1984; pp 164–176.

241. Ikekawa, N. In *Sterols and Bile Acids;* Danielson, H.; Sjövall, J., Eds.; Elsevier: Amsterdam, 1985; pp 199–230.

242. Minale, L.; Persico, D.; Sodano, G. *Experientia,* **1979**, *35*, 296, and earlier papers.

243. Rosa, M. D.; Minale, L.; Sodano, G. *Experientia,* **1980**, *36*, 360.

244. For pertinent references, see Lederer, E. *Q. Rev. Chem. Soc.* **1969**, *23*, 453; Nes, W. R.; McKean, M. L. *Biochemistry of Steroids and Other Isopentenoids;* University Park: Baltimore, MD, 1977; Harrison, D. M. *Nat. Prod. Rep.* **1985**, *2*, 525.

245. Idler, D. R.; Wiseman, P. M.; Safe, L. M. *Steroids* **1970**, *16*, 451.

246. Schmitz, F. J.; Pattabhiraman, J. *J. Am. Chem. Soc.* **1970**, *92*, 6073.

247. Fattorusso, E.; Magno, S.; Mayol, L.; Santacroce, C.; Sica, D. *Tetrahedron* **1975**, *31*, 1715.

248. Mattia, C. A.; Mazzarella, L.; Puliti, R.; Sica, D.; Zollo, F. *Tetrahedron Lett.* **1978**, 3953.

249. Kokke, W. C. M. C.; Shoolery, J. N.; Fenical, W.; Djerassi, C. *J. Org. Chem.* **1984**, *49*, 3742.

250. Tam Ha, T. B.; Djerassi, C. *Tetrahedron Lett.* **1985**, *26*, 4031.

251. Proudfoot, J. R.; Li, X.; Djerassi, C. *J. Org. Chem.* **1985**, *50*, 2026.

252. Cho, J.-H.; Djerassi, C. *J. Chem. Soc. Perkin Trans. 1* **1987**, 1307.

253. Steiner, E.; Djerassi, C.; Fattorusso, E.; Magno, S.; Mayol, L.; Santacroce, C.; Sica, D. *Helv. Chim. Acta* **1977**, *60*, 475.

254. Itoh, T.; Djerassi, C. *J. Am. Chem. Soc.* **1983**, *105*, 4407.

255. Barrow, K. D. In *Marine Natural Products*; Scheuer, P. J., Ed.; Academic Press: New York, 1983; Vol. 5, Chapter 2.

256. Stoilov, I. L.; Thompson, J. E.; Cho, J.-H.; Djerassi, C. *J. Am. Chem. Soc.* **1986**, *108*, 8235.

257. Catalan, C. A. N.; Thompson, J. E.; Kokke, W. C. M. C.; Djerassi, C. *Tetrahedron* **1985**, *41*, 1073.

258. Stoilov, I. L.; Thompson, J. E.; Djerassi, C. *Tetrahedron Lett.* **1986**, *27*, 4821.

259. Kikuchi, T.; Kadota, S.; Shima, T. *Tetrahedron Lett.* **1985**, 3817.

260. Li, L. N.; Sjöstrand, U.; Djerassi, C. *J. Am. Chem. Soc.* **1981**, *103*, 115.

261. Li, L. N.; Djerassi, C. *Tetrahedron Lett.* **1981**, *22*, 4639.

262. Khalil, M. W.; Djerassi, C.; Sica, D. *Steroids* **1980**, *35*, 707.

263. Theobald, N.; Djerassi, C. *Tetrahedron Lett.* **1978**, 4369.

264. Proudfoot, J. R.; Catalan, C. A. N.; Djerassi, C.; Sica, D.; Sodano, G. *Tetrahedron Lett.* **1986**, *27*, 423. Doss, G. A.; Proudfoot, J. R.; Silva, C. J.; Djerassi, C. *J. Am. Chem. Soc.* **1990**, *112*, 305.

265. Proudfoot, J. R.; Djerassi, C. *J. Chem. Soc. Perkin Trans. 1* **1987**, 1283.

266. Margot, C.; Catalan, C. A. N.; Proudfoot, J. R.; Sodano, G.; Sica, D.; Djerassi, C. *J. Chem. Soc. Chem. Commun.* **1987**, *19*, 1441. Doss, G. A.; Margot, C.; Sodano, G.; Djerassi, C. *Tetrahedron Lett.* **1988**, 6051. Doss, G. A.; Silva, J.; Djerassi, C. *Tetrahedron* **1989**, *45*, 1273.

267. Itoh, T.; Sica, D.; Djerassi, C. *J. Org. Chem.* **1983**, *48*, 890.

268. Li, L. N.; Li, H.; Lang, R. W.; Itoh, T.; Sica, D.; Djerassi, C. *J. Am. Chem. Soc.* **1982**, *104*, 6716.

269. Carlson, R. M. K.; Tarchini, C.; Djerassi, C. In *Frontiers of Bioorganic Chemistry and Molecular Biology*; Ananchenko, S. N., Ed.; Oxford: Pergamon, 1980; pp 211–224.

270. Ness, W. R. *Lipids* **1974**, *9*, 596.

271. Bloch, K. E. *CRC Crit. Rev. Biochem.* **1983**, *14*, 47; Yeagle, P. L. *Biochim. Biophys. Acta* **1985**, *822*, 267.

272. Craig, I. F.; Boyd, G. S.; Suckling, K. E. *Biochim. Biophys. Acta* **1978**, *508*, 418.

273. Walkup, R. D.; Jamieson, G. C.; Ratcliff, M. R.; Djerassi, C. *Lipids* **1981**, *16*, 631.

274. Litchfield, C.; Tyszkiewicz, J.; Dato, V. *Lipids* **1980**, *15*, 200; Morales, R. W.; Litchfield, C. *Biochim. Biophys. Acta* **1976**, *431*, 206; Litchfield, C.; Marcantonio, E. E. *Lipids* **1978**, *13*, 199.

275. Litchfield, C.; Greenberg, A. J.; Noto, G.; Morales, R. W. *Lipids* **1976**, *11*, 567; Morales R. W.; Litchfield, C. ibid. **1977**, *12*, 570.

276. Ayanoglu, E.; Walkup, R. D.; Sica, D.; Djerassi, C. *Lipids* **1982**, *17*, 617.

277. Ayanoglu, E.; Kornprobst, J. M.; Aboud-Bichara, A.; Djerassi, C. *Tetrahedron Lett.* **1983**, *24*, 1111; Ayanoglu, E.; Popov, S.; Kornprobst, J. M.; Aboud-Bichara, A.; Djerassi, C. *Lipids* **1983**, *18*, 830.

278. Wijekoon, W. M. D.; Ayanoglu, E.; Djerassi, C. *Tetrahedron Lett.* **1984**, *25*, 3285.

279. Dasgupta, A.; Ayanoglu, E.; Djerassi, C. *Lipids* **1984**, *19*, 768.

280. Kaneda, T. *J. Biol. Chem.* **1963**, *238*, 1229; idem. *Bacteriol. Rev.* **1977**, *41*, 391.

281. Carballeira, N.; Thompson, J. E.; Ayanoglu, E.; Djerassi, C. *J. Org. Chem.* **1986**, *51*, 2751.

282. Raederstorff, D.; Shu, A. Y. L.; Thompson, J. E.; Djerassi, C. *J. Org. Chem.* **1987**, *52*, 2337.

283. (a) Hahn, S.; Stoilov, I. L.; Ha, T. B. T.; Raederstorff, D.; Doss, G. A.; Li, H.-T.; Djerassi, C. *J. Am. Chem. Soc.* **1988**, *110*, 8117. (b) Hahn, S.; Lam, W.-K.; Wu, I.; Silva, C. J.; Djerassi, C. *J. Biol. Chem.* **1989**, *264*, 21043.

284. Jeffcoat, R.; James, A. T. In *Fatty Acid Metabolism and Its Regulation*; Numa, S., Ed.; Elsevier: Amsterdam, 1984; Chapter 4; Cook, H. W. In *Biochemistry of Lipids and Membranes*; Vance, D. E.; Vance, J. E., Eds.; Benjamin/Cummings: Menlo Park, Ca, 1985; Chapter 6.

285. Dasgupta, A.; Ayanoglu, E.; Wegmann-Szente, A.; Tomer, K. B.; Djerassi, C. *Chem. Phys. Lipids* **1986**, *41*, 335; Dasgupta, A.; Ayanoglu, E.; Tomer, K. B.; Djerassi, C. ibid. **1987**, *43*, 101.

286. Ayanoglu, E.; Wegmann, A.; Pilet, O.; Marbury, G. D.; Hass, J. R.; Djerassi, C. *J. Am. Chem. Soc.* **1984**, *106*, 5246.

287. Patton, G. M.; Fasulo, J. M.; Robins, S. J. *J. Lipid Res.* **1982,** *23*, 190.

288. Mena, P. L.; Pilet, O.; Djerassi, C. *J. Org. Chem.* **1984,** *49*, 3260.

289. Ayanoglu, E.; Li, H.-T.; Djerassi, C.; Düzgünes, N. *Chem. Phys. Lipids* **1988,** *47*, 165. Li, H.-T.; Düzgünes, N.; Ayanoglu, E.; Djerassi, C. ibid. **1988,** *48*, 109.

290. Mena, P. L.; Djerassi, C. *Chem. Phys. Lipids* **1985,** *37*, 257.

291. Ayanoglu, E.; Düzgünes, N.; Wijekoon, W. M. D.; Djerassi, C. *Biochim. Biophys. Acta* **1986,** *863*, 110.

292. Lawson, M. P.; Thompson, J. E.; Djerassi, C. *Lipids* **1988,** *23*, 741, 1037. Zimmerman, M. P.; Thomas, F. C.; Thompson, J. E.; Djerassi, C.; Streiner, H.; Evans, E.; Murphy, P. ibid. **1989,** *24*, 210.

293. Ayanoglu, E.; Rizzolio, M.; Beaulieu, S.; Roberts, J.; Oz, D.; Djerassi, C. *Comp. Biochem. Physiol.* **1990,** in press.

294. Djerassi, C.; George, T.; Finch, N.; Lodish, H. F.; Budzikiewicz, H.; Gilbert, B. *J. Am. Chem. Soc.* **1962,** *84*, 1499.

295. Djerassi, C. *Science* **1970,** *169*, 941.

296. *See also* Djerassi, C. *Science* **1969,** *166*, 468.

297. Djerassi, C. "The Future of Steroids in Contraception" *Wien. Med. Wochschr.* **1987,** *137*, 492.

298. Djerassi, C. In *Contraception in the Year 2001;* de Wied, D., Ed.; Amsterdam: Excerpta Medica Foundation, 1987; Chapter 13.

299. Djerassi, C. "The Bitter Pill" *Science* **1989,** *245*, 356.

300. Djerassi, C. "Fertility Awareness: Jet-Age Rhythm Method?" *Science* **1990,** *248*, 1061–1062.

301. Djerassi, C. "Abortion in the United States: Politics or Policy?" *Bull. At. Sci.* **1986,** *42*, 38.

302. Djerassi, C.; Israel, A.; Jöchle, W. *Bull. At. Sci.* **1973,** *29*, 10.

303. Djerassi, C.; Shih-Coleman, C.; Diekman, J. *Science* **1974,** *186*, 596.

304. Anon. *Chem. Eng. News* **Jan. 5, 1970,** pp 32–34; The Brazil Chemistry Program: An International Experiment in Science Education; American Chemical Society: Washington, DC, 1979.

305. Djerassi, C. *Bull. At. Sci.* **1968,** *24*, 23.

306. Voss, J. *Bull. Am. Acad. Arts Sci.* **1986,** *39*, 9; Rabinowitch, V. *Bull. At. Sci.* **1985,** *41*, 55; idem. *BOSTID DEVELOPMENTS;* National Research Council: Washington, DC, 1985; Vol. 5, p 1.

307. Crabbé, P.; Diczfalusy, E.; Djerassi, C. *Science* **1980**, *209*, 992. Crabbé, P.; Archer, S.; Benagiano, G.; Diczfalusy, E.; Djerassi, C.; Fried, J.; Higuchi, T. *Steroids* **1983**, *41*, 243.

308. Djerassi, C. *The Kenyon Review* **1986**, *8*, 91, 94.

309. Djerassi, C. *FRANK (Paris)* **1988**, *No. 10*, 90–100.

310. Djerassi, C. *The Futurist and Other Stories*; Macdonald–Futura: London, 1988.

311. Djerassi, C. *Cantor's Dilemma*; Doubleday: New York, 1989; Macdonald: London, 1990.

312. *Current Contents* Nov. 20, **1989**, *29*, 3.

313. Djerassi, C. *Cumberland Poetry Rev.* **1984**, *III(2)*, 63.

314. Djerassi, C. *Wallace Stevens J.* **1984**, *VIII(2)*, 110.

315. Djerassi, C. *Midwest Quart.* **1987**, *XXVIII(2)*, 214.

316. Djerassi, C. *New Lett.* **1987**, *54(1)*, 113.

317. Djerassi, C. *Negative Capability*; **1989**, *IX(1)*, 65.

318. Djerassi, C. *The Pill, Pygmy Chimps, and Degas's Horse*; MacMillan: New York, 1991.

319. Djerassi, C. *Hudson Rev.* **1989**, *XLII(1)*, 65.

320. Djerassi, C. *Hudson Rev.* **1989**, *XLII(1)*, 61.

321. Djerassi, C. *Negative Capability*; **1989**, *IX(1)*, 122.

322. Djerassi, C. *New Lett.* **1987**, *55(2)*, 13.

323. Djerassi, C. *Grand Street* **1988**, *8*, 167.

Index

A

Absolute configuration, determination by ORD, 54–59
Academic positions
  Stanford University, 66, 67
  Wayne University, 52–53
Activities in Third World countries, 145–151
Adams, Roger, 24 (photo), 79 (photo)
S-Adenosylmethionine, use in biomethylation, 123–125
Alkaloids
  personal experiment with mescaline, 14–17
  structural elucidations, 14
1,2-Alkyl rearrangements, demonstration by MS, 97
Allen, Charles, Jr., 103 (photo)
ALZA, 104
Amyrin, mass spectrometry, 80–81, 82
Androgens, partial aromatization to estrogens, 17–28
Animal sterols, general structure, 118
Antihistamines, studies at CIBA, 18, 22
Arbatov, Georgi, 148 (photo)
Arigoni, Duilio, 11 (photo), 160 (photo)
Aromatization of polycyclic molecules, studies by H. H. Inhoffen, 20–22
Arrival in the United States, 4
Aryl rearrangements, demonstration by MS, 97
Autobiographical poems, 158
Awakening (feminist group in Taiwan), members, 144 (photo)
Awards
  honorary doctorates
    Columbia University, 43
    UNAM, 146
  National Medal of Science, 40–41

Ayanoglu, Eser, unusual sterols, 129

B

Baldeschwieler, John, involvement in U.S.–Brazil chemistry program, 148
Barth, Günter, magnetic circular dichroism, 101, 102
Barton, D. H. R.
  at a Gordon Conference, 45 (photo)
  in Seoul, 153 (photo)
  molecular rotation differences, 54
  professional friend, 152
Batres, Enrique, cortisone synthesis from a plant source, 34 (photo)
Bendas, H., postdoctoral fellow, 58 (photo)
Berlin, J., cortisone synthesis from a plant source, 34 (photo)
Berson, Jerry, 11 (photo)
Bhutan journey, 161–162
Biomethylation of sterols, 123–125
Birch, Arthur J., 47 (photo), 151 (photo)
Bo, C. R., Taiwanese feminist, 144 (photo)
Board on Science and Technology in International Development (BOSTID), 150
Bowers, Albert, Syntex CEO, 58 (photo), 63
Brauman, John I., retro-Diels–Alder reaction, 84
Breslow, Ronald, 160 (photo)
Briat, Bernard, magnetic circular dichroism, 100
Brown, Keith
  at work at Stanford, 38 (photo)
  Brazilian natural products, 147

Brown, Peter, collaborator on review of electron-impact-induced rearrangements, 97, 100
Buchanan, Bruce, computer-aided interpretation of mass spectra, 107
Buchs, Armand, computer-aided interpretation of mass spectra, 107
Budzikiewicz, Herbert
  collaborator on mass spectrometry books, 80
  computer-aided interpretation of mass spectra, 107
  mass spectrometry of steroid ketones, 77, 79–80
Bunnenberg, Edward
  construction of low-temperature CD cell, 70
  magnetic circular dichroism, 100, 102
  optical circular dichroism, 67
Burlingame, Al, 86 (photo)
Burstein, Sumner, 61 (photo)
Bush, I., 45 (photo)

C

Cactus alkaloids, 13–14
Cancer operation, 162
Cantor's Dilemma (a novel), 160
Carballeira, Nestor, branching in long-chain fatty acids, 135
Carhart, Ray, expansion of DENDRAL to NMR spectrometry, 109
Carl XVI Gustaf, King of Sweden, 105 (photo)
Carothers, Wallace, Tarkio College alumnus, 6
Cetus Corporation, position held, 154
C_6H_6, stereochemical challenges, 112–113
Charge localization
  failure with saturated hydrocarbons, 92
  study by mass spectrometry, 90–92
Chemical ionization desorption mass spectrometry, 102
Cheng, C. H., Taiwanese feminist, 144 (photo)
Childhood education, 3
Chiral chromophoric derivatives
  concept, 69–70
  determination of rotamer composition, 71–72
  lecture topic, 70
  utility, 69
Chiral probes, use for conformational studies, 75–77
trans-2-Chloro-5-methylcyclohexanone, conformational analysis using octant rule, 62–63
Cholestane, mass spectra, 93f
Cholesterol, biological functions, 128
Chrysene, conversion into a polycyclic dienone, 22
CIBA Pharmaceutical Products, Inc., first industrial position, 18, 22–24
Circular dichroism (CD)
  analysis of monodeuterated cyclohexanones, 73–74
  early studies, 67–69
  principal limitation, 100
  temperature-dependent techniques, 70
  use in study of chiroptical behavior of molecules that owe chirality to isotopic substitution, 73–74
  variable-temperature CD for conformational analysis, 70–72, 74–75
Codisterol, use in sponge biosynthetic studies, 124–126
Cohn, Roy, Djerassi's surgeon, 114
College education, 4, 5, 18
Colton, Frank B., synthesis of norethynodrel, 50
Computer artificial intelligence applications, 102–114
  use in mass spectrometry, 106–109
  use in NMR spectral interpretation, 113–114
  use in structure elucidation, 109–114
Computer programs for structure elucidation
  CONGEN, 109
  DENDRAL, 105–109
  GENOA, 111–112
  use for stereochemical problems, 112–113

Computer programs for structure generation, STEREO, 112–113
Conformational analysis
  3-deuteriocyclohexanone, 73–75
  isotope effects, 73–77
  20-ketopregnanes, 70–71
  rotamers, 71–72
  use of chiral probes, 75–77
  use of variable-temperature CD, 70–72, 74–75
Conformationally immobile systems, examples, 76
CONGEN (constrained structure generation)
  pedagogic value, 110
  proposal to use as journal referee, 111
  use in structure elucidations, 109
Contraceptives, See Oral contraceptives
Coolidge, Walter H., chemistry teacher, 9
Corey, E. J., 160 (photo)
Cortisone
  first synthesis from a plant source, 32, 34 (photo)
  generation of ring-A substitution pattern, 30
  international race for its synthesis, 32, 34
  introduction of oxygen function, 30, 32–33, 40
  synthesis from steroidal sapogenins, 28–40
  synthesis from various starting materials, 34
  synthetic studies at CIBA, 23
Crabbé, P., postdoctoral fellow, 63
Crandell, Christopher, applications of computer techniques to NMR spectral interpretation, 113
Cyclohexanones, prediction of sign of ORD Cotton effect by octant rule, 60f
Cyclopropane ring
  formation in sterols, 126–127
  occurrence in sterols, 119–121
Cyclopropene, biosynthetic precursor, 126

D

Dawson, John, magnetic circular dichroism of biochemically important, optically active molecules, 101
de Kruif, Paul, influence on Djerassi, 9
25-Dehydroaplysterol, use in sponge biosynthetic studies, 124–125
Demospongic acids
  biosynthesis, 129, 135–136
  branching, 135
  plasma membrane constituents, 137
  proposed biogenetic pathways, 129, 135–136
  unusual features, 128–129, 130t–131t
DENDRAL (dendritic algorithm)
  procedure for interpretation of unknown mass spectrum, 107–108
  use in structure elucidation, 105–109
Desmotroposantonin, formation from santonin, 20–22
Deulofeu, Venancio, 146 (photo)
3-Deuteriocyclohexanone, conformational analysis, 73–75
Deuterium labeling
  studies of functional-group rearrangements, 97–100
  studies with steroid hydrocarbons, 94–95
  use in mass spectrometry, 86–95
Dienone–phenol rearrangement
  further studies, 26–28
  Ph.D. thesis, 22
Dihydrocalysterol, biosynthetic precursor of cyclopropane, 126–127
Diosgenin
  conversion to cortisone
    generation of ring-A substitution pattern, 30
    introduction of oxygen function, 30, 32
  Marker degradation, 28
  stages, 28
  source, 28

Djerassi, Alexander Maxwell
    (grandchild), 155, 159 (photo)
Djerassi, Alice (mother), 3, 4, 5 (photo)
Djerassi, Carl (photos)
    aboard a boat in Papua
        New Guinea, 139
    after a Japanese bath, 155
    as a graduate student, 10
    at a Columbia University
        commencement, 42
    at a Gordon Conference, 45
    at a Pugwash Conference, 148
    at an ICIPE meeting, 149
    at announcement of first cortisone
        synthesis, 34
    at birthday celebration, 152
    at birthday party of
        Ed Feigenbaum, 106
    at book-signing event, 161
    at IUPAC meeting in Japan, 85
    at IUPAC meeting in Prague, 39
    at Kurt Mislow's retirement party, 11
    at opening of Syntex Institute of
        Molecular Biology, 103
    at 1988 reunion, 160
    at Robert Maxwell's 60th
        birthday, 158
    at work on cortisone synthesis, 62
    clowning with lobster remnants, 124
    demonstrating skiing technique for
        stiff-legged persons, 66
    discussing circular dichroism, 71
    dressed in western garb, 8
    drinking alcohol-free beer, 16
    driving a tractor, 7
    during opening of Stauffer Organic
        Chemistry Building, 68
    in 1989 (after completing his
        autobiographical memoir), 170
    in Brazil, 146
    in Lucerne, 158
    in Nairobi, 150
    in research laboratory at Syntex, 26
    in Seoul, 153
    in Vladivostok, 163
    induction into National Inventors
        Hall of Fame, 51
    machete duel with Stork, 12
    magic duel with Nakanishi, 156

Djerassi, Carl (photos)—*Continued*
    mescaline party, 15
    on crutches, 64
    relaxing on a skiing trip with
        students, 137
    trekking in Bhutan, 162
    watching graduate students at
        Stanford, 38
    with boy scout troops on a skiing
        trip, 19
    with Carl Lenk, 55
    with choreographer Victoria Morgan
        and dancer Eda Holmes, 167
    with father, 4, 159
    with Günther Snatzke, 75
    with Huang Liang, 54
    with King Carl of Sweden, 105
    with Lord Todd, 78
    with members of a Taiwanese
        feminist organization, 144
    with mother, 5
    with opera star, 168
    with Paul Meier, 6
    with research group in 1955, 58
    with research group in 1957, 61
    with Robert Maxwell, 158
    with Russell Marker, 24
    with son Dale, 122, 159
    with Syntex consultants, 151
    with three generations of Djerassi
        males, 159
    with wife Diane Middlebrook, 164
Djerassi, Dale (son), 122 (photo), 155,
    159 (photo), 165 (photo)
Djerassi, Norma (second wife),
    169 (photo)
Djerassi, Pamela (daughter),
    165 (photo)
Djerassi, Samuel (father), 3, 4 (photo),
    152, 159 (photo)
Djerassi, Virginia (first wife),
    169 (photo)
Djerassi Foundation Artists Colony,
    160, 165, 166 (photo)
Donovan, F., postdoctoral fellow,
    58 (photo)
Dreiding, Andre
    in Lucerne, 158 (photo)
    personal and professional friend, 153

Dreiding, Norma, 158 (photo)
Düzgünes, Nezat, model membrane studies, 136

E

Education
  college education, 4, 5
  graduate studies, 19
  influential teachers, 9
  Newark Junior College, 4
  Realgymnasium, 3
  request for college scholarship, 5
  Tarkio College, 5
  undergraduate research, 9
Egli, Huldrych, applications of computer techniques to NMR spectral interpretation, 114
Ehrenstein, M.
  at a Gordon Conference, 45 (photo)
  transformation of strophanthidin, 43
Eisenbraun, E. J., collaborator from Wayne University, 67
Eisenbraun, Pete, 61 (photo)
El Supremo, title conferred by research fellow, 147
Electron-impact-induced rearrangements, 97–100
Element mapping, 97
Engle, Robert, 61 (photo)
Epiclerosterol, use in sponge biosynthetic studies, 125
Epicodisterol, use in sponge biosynthetic studies, 124–126
Equilenin, total synthesis, 19
Equilin, synthesis, 26–27
Estradiol, significance of formation from testosterone, 19–20
Estrogens
  by partial aromatization of androgens, 17–28
  computer-aided mass spectral analysis, 109
  synthesis from testosterone, 24–26
Estrone methyl ether, mass spectrometry, 80–81
Ethisterone
  activity, 47, 50
  structure, 48

Ethylene ketals
  charge localization, 90–92
  mass spectrometry, 90–92
17α-Ethynyltestosterone
  activity, 47, 50
  structure, 48
19-nor-17α-Ethynyltestosterone
  activity, 50
  patent application, 50
  preparation from 19-nortestosterone, 49, 50
Experimental biosynthesis
  practical problems, 123
  precursor incorporation, 123, 124

F

Faraday effect, 100
Fatty acids
  branching, 135
  demospongic acid types, 128, 130t–131t
  unusual types from sponges, 128–129, 130t–131t, 132t–133t
Feigenbaum, Edward
  at 50th birthday party, 106 (photo)
  implementation of DENDRAL, 105
  professional friend, 154
Fieser, Louis, 32 (photo)
Fieser, Mary, 32 (photo)
Figdor, Sandy, collaborator in mescaline experiment, 14
Fishman, Jack, 58 (photo)
Functional-group rearrangements, 97–100

G

Gallbladder operation, 114
Gallstones, mass spectral analysis, 115
Geller, Larry, 61 (photo)
GENOA, use in structure elucidation, 111–112
Gilbert, Ben, 61 (photo), 146 (photo), 147
Glaser, Donald, professional friend, 154

Goldbeck, Robert, magnetic circular dichroism of porphyrins, 102
Gordon Conference on Steroids and Related Natural Products, participants, 45 (photo)
Gorgosterol, structure elucidation, 115–116
Gorman, Marvin, collaborator in mescaline experiment, 14
Graduate student's wives, assistance in manual measurements, 54
Graduate studies
 confirmation of partial synthesis of an estrogen, 22
 fellowship, 19
 part-time study, 19
 Ph.D. thesis topic, 19–20, 22
 University of Wisconsin, 10
Gray, Harry, involvement in U.S.–Brazil chemistry program, 148
Gray, Jim, 61 (photo)
Gray, Neil, applications of computer techniques to NMR spectral interpretation, 113
Green, Mark, interaction of functional groups during mass spectrometry, 97
Grossman, J., postdoctoral fellow, 58 (photo)
Gunther, John, plagiarizing by Djerassi, 6–8

H

Haberlandt, Ludwig, biological functions of progesterone, 43
Hahn, Soonkap, demospongic acid biosynthesis, 135
Halogenated cyclohexanones, correlation between UV absorption and ORD, 67–68
α-Haloketone effect, 59
α-Haloketone rule, relation to octant rule, 59
Halpern, Otto, 61 (photo), 63
Hammond, George, involvement in U.S.–Brazil chemistry program, 148

Hecogenin
 structure, 35
 use in synthesis of cortisone, 34
Henry, J. A., postdoctoral fellow, 58 (photo)
Heusser, K., at a Gordon Conference, 45 (photo)
Himalayan journey, 161–162
Hodges, R., 58 (photo)
Hoffman, Frances, 42 (photo)
Holmes, Eda, dancer, 167 (photo)
Honorary doctorates, 43, 146
Huang, S. C., Taiwanese feminist, 144 (photo)
Huttrer, Charles, collaboration on antihistamines, 18

I

Industrial positions
 ALZA, 104
 Cetus Corporation, 154
 CIBA Pharmaceutical Products, Inc., 18, 22–24
 Syntex, 23, 63
 SYVA, 104
 Teknowledge, 154
 Zoecon Corporation, 104, 145
Inhoffen, H. H., selective aromatization of polycyclic molecules, 20–22
Inside Europe, plagiarizing by Djerassi, 6–8
Institute for International Education, scholarship assistance, 5
Instituto de Quimica Agricola (Brazil), 147
Instituto de Quimica of the National University of Mexico (UNAM), 146
International Centre for Insect Physiology and Ecology (ICIPE), 149–150
INTSUM, computer program for interpretation of mass spectral fragmentations, 109
Iodoketones, correlation between UV absorption and ORD, 67–68
Iresin
 degradation products, 57–58

Iresin—*Continued*
  determination of absolute
    configuration, 56
Iriarte, Jose, Syntex collaborator, 24
Isoprene rule, 56
Isotope effects in conformational
  isomerism, 73–77
Isotope labeling, use in mass
  spectrometry, 85–87
Itoh, Toshihiro, separation and
  identification of sponge
  sterols, 116

J

Jeger, O., at a Gordon Conference,
  45 (photo)
Johnson, William S.
  at a Gordon Conference, 45 (photo)
  during opening of Stauffer Organic
    Chemistry Building, 68 (photo)
  invitation of Djerassi
    to Stanford, 66–67
  involvement in U.S.–Brazil
    chemistry program, 148
  professional friend, 155
Jones, E. R. H.
  at a Gordon Conference, 45 (photo)
  during opening of Stauffer Organic
    Chemistry Building, 68 (photo)

K

Kan, Peter, 61 (photo)
Kapoor, Ami, 61 (photo)
Karliner, Jerry, sterol mass
  spectrometry, 115
Kenyon College, 5, 18
Ketones, isotopically engendered
  chirality, 73–74
20-Ketopregnanes, conformational
  analysis, 70–71
Klyne, W., molecular rotation
  differences, 54
Knee surgery, 63
Knowles, Jeremy, 11 (photo)

Krakower, Gerald, 58 (photo),
  61 (photo)
Kreps, Juanita (Secretary of
  Commerce), 51 (photo)
Kupperman, Aaron, involvement in
  U.S.–Brazil chemistry
  program, 148
Kutney, Jim, 61 (photo), 63

L

Laboratorios Syntex S.A.
  laboratory equipment, 23–24
  research laboratory, 26 (photo)
  technical director, 23
  See also Syntex
Lectures
  Centenary Lecture of the Royal
    Chemical Society, 70
  first honorarium, 7–9
  1982 IUPAC lecture, 106
  Second International Congress on
    Hormonal Steroids (plenary
    lecture), 115
Lederberg, Joshua
  at opening of Syntex Institute of
    Molecular Biology, 103 (photo)
  interest in artificial intelligence,
    104–105
  involvement in mass
    spectrometry, 106
  professional relationship with
    Djerassi, 102–103
Lee, Y. C., Taiwanese feminist,
  144 (photo)
Lehmann, Pedro, 24 (photo)
Lehn, Jean-Marie, 160 (photo)
Lemming, Peter, 61 (photo)
Lenk, Carl, first Ph.D. student,
  55 (photo)
Liang, Huang, postdoctoral fellow,
  54 (photo)
Lightner, David A.
  at work at Stanford, 38 (photo)
  refinements to octant rule, 62
  syntheses of conformationally
    immobile systems, 76

Limbach, H., magnetic circular dichroism of porphyrins, 102
Linder, Robert, magnetic circular dichroism, 101, 102
Ling, Nicholas, X-ray analysis of gorgosterol, 115–116
Literary work
  autobiographical poems, 158
  autobiography, 165
  memoirs, 166
  novels, 160–161, 165
  peers as inspiration for fictional heroes, 155, 157
  poetry, 165
  short stories, 165
Lophenol, structure elucidation, 77
Lowell, Robert, poet, 168 (photo)

M

Macrolide antibiotics, 13
Magic duel with Koji Nakanishi, 156–157
Magnetic circular dichroism
  porphyrins, 101–102
  study of biochemically important, optically active molecules, 101
Magnetic optical rotation, 100
Mancera, Octavio, Syntex collaborator, 24, 34 (photo)
Marine sterols
  biomethylation, 123–125
  biosynthesis of sponge sterols, 122–129
  occurrence of cyclopropane ring, 119–121
  possible biological role, 128
  structural variability, 118–121
  structure elucidation, 116–121
  See also Sponge sterols
Marker, Russell, 24 (photo)
Marker degradation, in conversion of diosgenin to cortisone, 28–29
Mass spectrometry
  amyrin, 80–81, 82
  cholestane, 93*f*

Mass spectrometry—*Continued*
  computer-aided interpretation of spectra, 106–109
  determination of charge localization, 90–92
  Djerassi's gallstones, 115
  element mapping, 97
  estrogenic hormones, 109
  estrone methyl ether, 80–81
  ethylene ketals, 90–92
  functional-group rearrangements, 97–100
  gorgosterol, 115
  isotope-labeling studies, 85–87
  4-methoxycyclohexanone, 97, 99
  occurrence of functional-group rearrangements, 97
  octalone, 97–98
  phospholipids, 136
  porphyrins, 102
  publications, 79–80, 84
  saturated steroid hydrocarbons, 92–95
  stelliferasterol, 95–96
  steroid ketones, 77, 79–80, 87–92
  steroidal ketals, 90–92
  taraxerols, 80–81, 82
  triterpenes, 80–81, 82
  unsaturated steroid hydrocarbons, 95–96
  α,β-unsaturated steroid ketones, 88–90
  use in structure elucidation, 80–84
  use of deuterium labeling, 86–95
  vincadifformine, 81–83
Mauli, Rolf, 61 (photo), 63
Maxwell, Isabel (daughter-in-law), 155
Maxwell, Robert
  birthday (60th), 154 (photo)
  personal and professional relations, 155
  poetry subject, 155
Mayer, Rudolf L., collaborator on antihistamines, 18
McLafferty, Fred, 86 (photo)
McLafferty rearrangement, 87–88, 96
Meier, August, 4
Meier, Clara, 4

Meier, Frank, 4
Meier, Paul, 4, 6 (photo)
Meinwald, Jerrold
   in Nairobi, 150 (photo)
Meinwald, Jerrold—*Continued*
   involvement in ICIPE, 150
Mescaline
   effects, 15
   effects on Djerassi, 16–17
   personal experiment, 14–17
Mescaline party, 15 (photo)
4-Methoxycyclohexanone, mass
   spectrum, 97, 99
Methyl migration, in steroid
   synthesis, 22
Methymycin, structure, 13
Meyers, Al, 11 (photo)
Microanalyses, uselessness according
   to Djerassi, 10–11
Microbe Hunters, influence on
   Djerassi, 9
Microbiological fermentation
   technology, industrial use
   in steroid synthesis, 40
Middlebrook, Diane (wife) (photos),
   139, 158, 164
Middlebrook, Leah (stepdaughter),
   161 (photo)
Miescher, K., CIBA director, 18
Mills, J. S., 63
Miramontes, Luis, collaborator in first
   synthesis of a synthetic oral
   contraceptive, 50
Mislow, Kurt
   collaboration with Djerassi, 69
   discussing circular dichroism,
      71 (photo)
   mathematical analysis of Djerassi
      publication output, 11–12
   ORD properties of hindered biaryls,
      68–69
   retirement party, 11 (photo)
Mitscher, Lester, 61 (photo)
Modeer, Bengt, 105 (photo)
Moffitt, William
   clowning with Woodward,
      60 (photo)
   octant rule, 59

Monodeuterated cyclohexanones, CD
   measurements, 73–74
Monomethylcholesterol derivatives,
   biosynthetic significance, 77
Monsimer, H. S., 58 (photo)
Monteiro, Hugo, successful wager
   against Djerassi, 147
Morgan, Victoria, choreographer,
   167 (photo)
Mors, Walter, collaborator in
   cooperative program for exchange
   of postdoctoral fellows, 147
Moscowitz, Albert
   at Kurt Mislow's retirement party,
      11 (photo)
   collaboration with Djerassi, 69
   discussing circular dichroism,
      71 (photo)
   effect of asymmetric solvation
      around a chromophore, 70
   inherently dissymmetric
      chromophores, 68
   magnetic circular dichroism, 101
   octant rule, 59
Murray, K. E., 79 (photo)

N

Nakanishi, Koji
   after a Japanese bath, 155 (photo)
   in Nairobi, 150 (photo)
   inspiration for fictional hero, 156
   involvement in ICIPE, 150
   magic duel with Djerassi, 156–157
   magic for melon, 154
   professional friend, 153–154
   with Syntex consultants, 151 (photo)
National Inventors Hall of Fame,
   induction of Djerassi, 50,
   51 (photo)
National Medal of Science award,
   encounter with Richard Nixon,
   40–42
Natural products chemists, lack of
   peer recognition, 157–158, 160
*Natural Products Related to Phenan-
   threne*, influence on Djerassi, 18

Neohamptogenin hoax
  draft of communication, 36
  neohamptogenin structure, 35
  presentation by Woodward
    at a Gordon Conference, 37–38
    at a Harvard seminar, 39
Neosynalar, development at Syntex, 65
Newark Junior College, 4
Nixon, Richard, 40, 41 (photo), 43
NMR spectrometry, computer-assisted analysis, 110–111, 113–114
Norethindrone
  biological activity, 50
  formation from norethynodrel, 50
  human trials, 50
  patent application, 50
  preparation from 19-nortestosterone, 49, 50
  trade name, 52
Norethynodrel
  activity, 52
  conversion to norethindrone, 50
  patent application, 50
Norgestrel, structure, 52
Norlutin, *See* Norethindrone
19-Norprogesterone
  activity, 44
  structure, 44, 48
  synthesis, 44, 48
19-Nortestosterone
  conversion to 19-nor-17α-ethynyltestosterone, 49, 50
  structure, 27, 46
  synthesis, 44, 46
Norton, Bayes M., chemistry teacher, 9
Nourse, James, solutions to Rubik's cube, 112
Nussbaum, A. L., cortisone synthesis from a plant source, 34 (photo)

O

Octalins, retro-Diels–Alder, 84
Octalone, mass spectrum, 97–98
Octant rule
  consequences, 61–62
  example of application, 62–63
  extension, 68–69
  original version, 59–60, 62
  prediction of sign of ORD Cotton effect of cyclohexanones, 60f
  refinements, 62
  relation to α-haloketone rule, 59
Odhiambo, Thomas, ICIPE, 149–150
Optical activity, natural vs. magnetically induced, 100–101
Optical circular dichroism, studies at Stanford, 67
Optical rotatory dispersion (ORD)
  in determination of absolute configuration, 54–59
  principal limitation, 100
Optical rotatory dispersion curves, diagnostic tool, 55–56
Oral contraceptives
  first synthesis, 40–53
  See also specific compounds
Organic mass spectrometry, *See* Mass spectrometry
Osiecki, Jeanne, 61 (photo)
Overberger, Charles, involvement in U.S.–Brazil chemistry program, 148

P

Pakrashi, S. C., 58 (photo)
Pataki, J., cortisone synthesis from a plant source, 34 (photo)
Patents, circumvention, 52
Peer recognition
  problem of natural products chemists, 157–158, 160
  subject of Djerassi novel, 160
Peng, W., Taiwanese feminist, 144 (photo)
Pentacyclic triterpenes, 13–14
Personal reflections
  Himalayan journey, 161–162
  mortality, 162
  nonchemical personae, 163–166
Petrosterol, biosynthesis, 126–127
Pettit, G. R., 58 (photo)

Phospholipid(s)
  characterization, localization, and liposome formation, 136–137, 142
  Djerassi's entry into the field, 128
  mass spectrometry, 136
  model membrane studies, 136
  studies, 128–137
Pilocereine, structure, 14
Pizza-*cum*-mescaline sulfate, menu for Sunday picnic, 15
Plant sterols
  general structure, 118
  vs. marine sterols, 118
Policy courses taught, 139, 144
Polycyclic molecules, selective aromatization, 20–22
Polygamous life, 9
Porphyrins
  magnetic circular dichroism, 101–102
  mass spectral investigations, 102
  synthetic work, 102
Postdoctoral fellows, 63, 65
Prelog, Vlado, 85 (photo)
Professional affiliations
  Brazilian Research Council, 147
  ICIPE, 149–150
  National Academy of Sciences, 147, 150
  World Health Organization, 150
Professional friends, 151–157
Progesterone
  biological functions, 43
  medical use, 43
  microbial oxidation, 40
  structural specificity of activity, 43
Public policy issues
  birth control, 145
  EPA registration policies, 145
Public speaking, early attempts, 6–9
Publications
  applications of mass spectrometry in the alkaloid field, 84
  artificial intelligence for chemical inference, 106
  first review article, 80
  impact of "A High Priority? Research Centers in Developing Nations", 148–149

Publications—*Continued*
  impact of "Birth Control after 1984", 145
  impact of "Insect Control of the Future: Operational and Policy Aspects", 145
  importance in early career, 23
  "Mass Spectrometry in Structural and Stereochemical Problems", 79
  mass spectrometry of organic chemical functionalities, 100
  "Natural Products Chemistry 1950 to 1980—A Personal View", 14
  novel marine sterols, 116–117
  organic mass spectrometry, 79–80
  output during Syntex period, 24
  preparation of drafts, 13
  productivity, 11–12
  review on element mapping, 97, 100
  role in scientific work, 13
  Steroid Reactions, 17
  translation into Japanese, 153
  unusual royalty contract with McGraw–Hill, 65
Pyribenzamine (tripelennamine), structure, 18

R

Raederstorff, Daniel, demospongic acid biosynthesis, 135
Research career
  beginnings, 10
  collaborators, 12–13
  diversity of interests, 9
  move from industry to academia, 23
  productivity in publications, 11–12
Research subjects
  alkaloids, 14
  antihistamines, 18
  at Stanford University, 67–69
  at Syntex, 24–26
  experimental biosynthesis, 122
  macrolide antibiotics, 13
  magnetic circular dichroism, 100–102
  marine sterols, 114–126
  organic mass spectrometry, 77–100

Research subjects—*Continued*
  pentacyclic triterpenes, 13–14
  phospholipids, 126, 128–129
  steroids, 17–53
  topics other than steroids, 13–17
Retro-Diels–Alder reaction
  octalins, 84
  Δ^7-steroid olefins, 84
  use in structure elucidation, 80–83
Rinehart, Ken, 86 (photo)
Ringold, Howard, 151 (photo)
Riniker, B., 58 (photo)
Roberts, J. D., 160 (photo)
Robertson, Alexander, computer-aided interpretation of mass spectra, 107
Robinson, Cecil, 58 (photo)
Robinson, Robert, 79 (photo)
Romo, Jesus, Syntex collaborator, 24, 34 (photo)
Roosevelt, Eleanor, scholarship assistance, 5
Rosengart, Siegfried, art dealer, 158 (photo)
Rosenkranz, George
  at Djerassi's 60th birthday, 152 (photo)
  cortisone synthesis from a plant source, 34 (photo)
  professional friend, 152
  Syntex technical director, 23
Rotamers, conformational analysis, 71–72
Rubik's cube, solutions by James Nourse, 112
Ruzicka, Leopold
  in Zurich, 57 (photo)
  isoprene rule, 56

S

Sahlberg, Karl-Erik, 105 (photo)
Sandberg, Finn, 153 (photo)
Santonin, conversion to desmotroposantonin, 20–22
Sapogenins, conversion to cortisone, 28–40

Sarett, L. H.
  at a Gordon Conference, 45 (photo)
  involvement in neohamptogenin hoax, 35–39
Scheuer, Paul, mass spectrometry of marine sterols, 115
Scholarship assistance, 5
Schooley, David, magnetic circular dichroism, 100
Schroll, Gustav, computer-aided interpretation of mass spectra, 107
Scientific work, Djerassi's criteria of completion, 13
Scuba diving, 123
Selective aromatization, polycyclic molecules, 20–22
Sheehan, John, move from industry to academia, 23
Shu, Arthur, total synthesis of a demospongic acid, 135
Sih, C. J., 151 (photo)
Skiing accident, 18
Smith, Dennis, application of artificial intelligence to mass spectrometry, 108–109
Smith, Howard, 61 (photo)
Snatzke, Günther, 75 (photo)
Sondheimer, Franz, 151 (photo)
Spectropolarimeter, in determination of absolute configuration, 55
Sponge fatty acids, unusual features, 128–129, 130*t*–131*t*
Sponge sterols
  biosynthesis, 122–129
  general structural features, 117–118
  variation in side chains, 117–119
  See also Marine sterols
Spoof research report, text written by Ben Tursch, 140–143
Staal, Gerardus, 105 (photo)
Stanford Industrial Park, 103, 104
Stauffer Chemical Laboratory (Stauffer Organic Chemistry Building), construction, 67
Stelliferasterol, mass spectral analysis, 95–96
STEREO, use in structure generation, 112–113

Stereochemical problems,
  computer-aided solutions, 112–113
Steroid(s)
  contribution to solution of
    mechanistic problems, 28
  halogenation–dehydrohalogenation,
    23, 24–26
  introduction to these compounds at
    CIBA, 18
  partial synthesis, 17–53
  studies carried out at Syntex, 24–26
  topic for graduate research, 22
  total synthesis, 19
Steroid hydrocarbons
  deuterium-labeling studies, 94–95
  mass spectra
    saturated hydrocarbons, 92–95
    unsaturated hydrocarbons, 95–96
Steroid ketones
  mass spectrometry, 77, 79–80, 87–92
  ORD experiments, 55–59
  structure and H–O distance, 88t
Δ^7-Steroid olefins, retro-Diels–Alder
  reaction, 84
Steroid Reactions, rationale for book,
  17–18
Steroidal ketals
  charge localization, 90–92
  mass spectrometry, 90–92
Sterols, biomethylation, 123–125
Stoilov, Ivan, collaborator in marine
  studies, 123, 129
Stork, Gilbert
  cortisone synthesis from a plant
    source, 34 (photo)
  flashbulb explosion in Mexico, 146
  in Paris, 37 (photo)
  inspiration for fictional hero,
    155–156
  involvement in neohamptogenin
    hoax, 35–39
  machete duel with Djerassi,
    12 (photo)
  professional friend, 151
  with Syntex consultants, 151 (photo)
Stronglyosterol, biosynthesis, 125–126
Strophanthidin, conversion to
  19-norprogesterone, 44

Structure elucidation
  gorgosterol, 115–116
  marine sterols, 116–121
  mechanistic rationalizations, 80
  tabersonine, 84
  use of computer artificial
    intelligence, 109–114
  use of mass spectrometry, 80–84
  use of retro-Diels–Alder reaction,
    80–83
Structure generation with overlapping
  atoms (GENOA), 111–112
Students
  Djerassi's chemistry progeny,
    138–139, 144
  move from Wayne University
    to Mexico City, 63, 65
  Wayne University, 58 (photo),
    61 (photo)
Synalar, development at Syntex, 65
Syntex
  Djerassi's research group, 63–65
  positions held, 24, 63
  research subjects, 65
  winner in race for cortisone
    synthesis, 32
  See also Laboratorios Syntex S.A.
Syntex Institute for Molecular
  Biology, 103, 104
SYVA, 104

T

Tabersonine, structure elucidation, 84
Takeda, Ken'ichi, 85 (photo)
Taraxerols, mass spectrometry,
  80–81, 82
Tarkio College, 5, 6
Tarkio Rotary Club, 6
Taube, Henry, involvement in
  U.S.–Brazil chemistry
  program, 148
Teaching, 138–139, 144
Teknowledge, position held, 154
Terman, Fred E., Stanford University
  provost, 67, 102, 104

Testosterone, significance of conversion to estradiol, 19–20
Thieberg, Kay, 61 (photo)
Third World affiliations
  BOSTID, 150
  ICIPE, 149–150
  Instituto de Quimica Agricola (Brazil), 147
  Instituto de Quimica of the National University of Mexico (UNAM), 146
  Syntex (Mexico), 145–146
  WHO Special Programme of Research, Development and Research Training in Human Reproduction, 150–151
Thompson, Janice, collaborator in marine studies, 123, 129
Todd, Lord, 78 (photo)
Topical corticosteroids, development at Syntex, 65
Torgov, I. V., at an IUPAC conference, 39 (photo)
Tripelennamine (pyribenzamine), structure, 18
Triterpenes, mass spectrometry, 80–81, 82
Troyanos, Tatiana, opera star, 168 (photo)
Trudell, J. R., magnetic circular dichroism of biochemically important, optically active molecules, 101
Tryptophan, development of quantitative determination by magnetic circular dichroism, 101
Tursch, Ben
  Brazilian natural products, 147
  on a boat in Papua New Guinea, 139 (photo)
  spoof research article, 138, 140–143
Tyler, Edwards, clinical results with norethindrone, 50

U

Undergraduate research, 9

University of Wisconsin Chemistry Department, facilities in 1942, 10
α,β-Unsaturated steroid ketones, mass spectral fragmentation, 88–90
Upjohn Company, microbiological fermentation technology, 40
U.S.–Brazil chemistry program, 147

V

Velasco, Mercedes, cortisone synthesis from a plant source, 34 (photo)
Villotti, Riccardo, 61 (photo), 63
Vincadifformine, mass spectrometry, 81–83
von Ragué Schleyer, Paul, 11 (photo)

W

Walkup, Robert, characterization of demospongic acid fraction, 128
Washton, Nathan, chemistry teacher, 9
Wayne State University, *See* Wayne University
Wayne University, Djerassi research group, 58 (photo), 61 (photo), 63
Wellman, Keith, construction of low-temperature CD cell, 70
Wenkert, Ernest, 146 (photo)
Wettstein, A., CIBA director, 18
"White House enemy list", 40
Wilds, A. L.
  modification of Birch reduction, 44, 47
  Ph.D. thesis advisor, 19
  photo, 19
  total synthesis of steroids, 19
Williams, Dudley H., collaborator on mass spectrometry books, 80
Woodward, R. B.
  at a Gordon Conference, 45 (photo)
  at an IUPAC conference, 39 (photo)
  clowning with Moffitt, 60 (photo)
  involvement in neohamptogenin hoax, 35–39
  personal qualities, 37

World Health Organization (WHO), 150
Wu, C. L., Taiwanese feminist, 144 (photo)
Wynberg, Hans, syntheses of conformationally immobile systems, 76

X

Xestosterol, use in sponge biosynthetic studies, 124–125

Y

Yashin, Rosa, cortisone synthesis from a plant source, 34 (photo)

Z

Zaffaroni, Alejandro
  at Djerassi's 60th birthday, 152 (photo)
  founder of ALZA, 104
  inspiration for fictional hero, 156
  professional friend, 152
  Syntex executive vice-president, 53 (photo)
Zderic, John, Syntex vice-president, 58 (photo), 65
Zen experience, 161–162
Zoecon Corporation, position held, 104, 145

Production: Paula M. Bérard *Copy Editing: Colleen P. Stamm*
Indexing: Ann Maureen R. Rouhi *Acquisition: Robin Giroux*

Printed and bound by Maple Press, York, PA

Paper meets minimum requirements of American National Standard for Information Sciences—Permanence of Paper for Printed Library Materials, ANSI Z39.48–1984 ∞

Profiles, Pathways, and Dreams

Sir Derek H. R. Barton *Some Recollections of Gap Jumping*

Arthur J. Birch *To See the Obvious*

Melvin Calvin *Following the Trail of Light: A Scientific Odyssey*

Donald J. Cram *From Design to Discovery*

Michael J. S. Dewar *A Semiempirical Life*

Carl Djerassi *Steroids Made It Possible*

Ernest L. Eliel *From Cologne to Chapel Hill*

Egbert Havinga *Enjoying Organic Chemisty 1927–1987*

Rolf Huisgen *Mechanisms, Novel Reactions, Synthetic Principles*

William S. Johnson *A Fifty-Year Love Affair with Organic Chemistry*

Raymond U. Lemieux *Explorations with Sugars: How Sweet It Was*

Herman Mark *From Small Organic Molecules to Large: A Century of Progress*

Bruce Merrifield *The Concept and Development of Solid-Phase Peptide Synthesis*

Teruaki Mukaiyama *To Catch the Interesting While Running*

Koji Nakanishi *A Wandering Natural Products Chemist*

Tetsuo Nozoe *Seventy Years in Organic Chemistry*

Vladimir Prelog *My 128 Semesters of Studies of Chemistry*

John D. Roberts *The Right Place at the Right Time*

Paul von Rague Schleyer *From the Ivy League into the Honey Pot*

F. G. A. Stone *Organometallic Chemistry*

Andrew Streitwieser, Jr. *A Lifetime of Synergy with Theory and Experiment*

Cheves Walling *Fifty Years of Free Radicals*

9780841217737.3